WATKINS MILL

Watkins MILL

THE FACTORY *on the* FARM

LOUIS W. POTTS &
ANN M. SLIGAR

TRUMAN STATE UNIVERSITY PRESS

tsup.truman.edu

Cover art: Watkins Mill and Bethany Home, from the *Illustrated Historical Atlas of Clay County, Missouri*
Cover designer: Teresa Wheeler
Type: Bookman Old Style™ by The Monotype Corporation; LegacySerITC © by URW
Printed by Thomson-Shore, Dexter, Michigan USA

Library of Congress Cataloging-in-Publication Data
Potts, Louis, 1944–
Watkins Mill : the factory on the farm / by Louis W. Potts, Ann M. Sligar.
p. cm.
Includes bibliographical references and index.
ISBN 1-931112-22-3 (alk. paper)
1. Watkins Woolen Mill State Historic Site (Lawson, Mo.) 2. Watkins, Waltus L., 1806–1884. 3. Watkins family. 4. Businessmen—Missouri—Lawson—Biography. 6. Wool industry—Missouri—Lawson—History—19th century. 7. Woolen and worsted manufacture—Missouri—Lawson—History—19th century. 8. Agriculture—Missouri—Lawson—History—19th century. 9. Lawson (Mo.)—Biography. 10. Lawson (Mo.)—Buildings, structures, etc.
I. Sligar, Ann M. II. Title.
F474.L3P68 2003
977.8'16—dc21

2002014248

♾ The paper in this publication meets or exceeds the minimum requirements of the American National Standard for Information Sciences–Permanence of Paper for Printed Library Materials, ANSI Z39.48-1992.

To the Watkins Mill Association's founders
and others who helped save the mill
and sustain its interpretation

The older parts of the country, which have been settled by the husbandman, and reclaimed from the state of nature, are now to be settled again by the manufacturer, the engineer, and the mechanic. First settled by a civilized, now they are to be settled by a dense, population. Settled by the hard labor of human hands, they are now to be settled by the labor-saving arts, by machinery, by the steam engine, and by internal improvements.

—*Edward Everett*

Contents

ILLUSTRATIONS

Color Insert:

Credits:

Jet Lowe .... Historic American Engineering Record (HAER, MO-24, Laws)
DNR .... Missouri Department of Natural Resources
WMA... Watkins Mill Association
WMSHS .... Watkins Mill State Historic Site, Lawson, Missouri

Figures:

PREFACE

On the western frontier of Missouri in 1839, an enterprising Kentucky emigrant named Waltus Watkins (1806–84) established a rude farm that would shortly evolve into a bustling community centered on his numerous commercial ventures. As wagon paths grew into a thoroughfare linking nearby county seats, Watkins's farmstead in Clay County became a place where neighboring farmers came to pick up their mail; where a continuous stream of kin or acquaintances from Tennessee and Kentucky paid visits; where Friday night meetings of the local debate society featured the orations of the young males in the three-county community.

Over the next two decades, Watkins developed a variety of ventures aimed at processing the yield of his flourishing agricultural pursuits. Neighbors from perhaps as far away as fifteen miles hauled their corn, logs, and livestock to his mills and barns. In turn flour, cornmeal, and lumber went home in the farm wagons or were sent to market. Then, on the eve of the Civil War, Waltus's most imposing and opportunistic enterprise—his woolen textile factory—was created. From farflung national networks of steamboats and railroads came dyestuffs, machinery, and tons of raw wool destined for Watkins Mill. Yarn, cloth, shawls, and blankets were dispatched to customers or consigned to storekeepers within a seventy-five-mile radius. Mill workers of all ages, genders, and skill levels were drawn from industrial centers in the eastern U.S. and Europe, as well as from nearby farms. Entire families were employed at the mill, extending Watkins's impact and influence to multiple generations.

At Watkins Mill, the habits and values associated with the rural South mingled with more modern behaviors and attitudes linked with industrial society. Rather than viewing these contrasting folkways as the source of conflict, Watkins sought to make them complementary, to blend the dynamism of the factory with his more progressive agrarian pursuits. He wanted his factory to remain in a rural setting, as a pastoral experiment. Watkins Mill was to be, in historian John Kasson's terms, "a middle distance, between city and wilderness." There would be no intense focus on industrialism here; no city or even town arose from this community.[1]

Watkins's Jeffersonian experiment flourished for a number of years, in spite of severe political and economic challenges. The combination of Watkins's death in 1884 and the rapid decline in the woolen industry in the Midwest at about the same time led to the mill's eventual failure. The woolen factory was for the most part silent for the last fifteen years of the century, and the community around it slowly dwindled. The farmstead itself remained vibrant, as the Watkins offspring who remained grew acres of oats and hay, tended herds of Shorthorn cattle and milk cows, and pruned extensive orchards, all (like their father) with an eye on the market.

The story of Watkins Mill is both archetypal—Waltus, his family, associates, and employees; his westward migration; his social values; and his entrepreneurial drive represent much of nineteenth-century America—and unusual—the extent of Watkins's enterprises and his purposeful and successful combination of agriculture and industry. Similarly, it was both a societal upsurge of interest in historic preservation and a remarkable local political effort that saved the mill and homestead for posterity.

To tell the story, this book weaves together many threads: family correspondence, business documents, church archives, newspapers, census records, archaeological evidence, folklore, oral history, and perhaps most important of all, the buildings and machines remaining at the site. Since 1964, Watkins Mill has been a Missouri State Historic Site; it is also designated as a National Historic Landmark and a National Historic Mechanical Engineering Landmark. In the words of industrial chronicler Laurence Gross, it "contains the finest collection of nineteenth-century textile machinery in situ in North America and includes the most significant textile artifacts known to survive.... The machinery and building can provide modern viewers a great deal of information about the cloth-making process and about the history of machine-building, industrial development, and the working life of the men and women who worked there." Although the factory might now be silent, the structure and the machines are not mute. Combined with the documentary records of the Watkinses' agricultural and industrial endeavors, they offer testimony of how the Industrial Revolution came to rural Missouri in the mid nineteenth century.[2]

Notes for Preface

[1]Kasson, *Civilizing the Machine,* 19–28.

[2]Gross, "Importance of Research," 16.

ACKNOWLEDGMENTS

Our scholarly quest has been a social experience emerging from family and moving outward in concentric circles. Louis's parents, Carl and Joyce, stimulated his imagination by introducing him to historical novels. His wife, Annetta, has sustained his career in manifold and selfless ways. Ann's family, in its frequent moves across America and in Europe, revealed the variety of landscape, artifact, and culture in our heritage.

Financial support from the American Association for State and Local History provided initial research ably performed by Plora Walby. The Missouri Humanities Council (the state affiliate of the National Endowment for the Humanities), with supplemental funding from the Watkins Mill Association, the Ferrell Companies, Inc., the Perry Foundation, and the Henry Wurst family sponsored production of an orientation video for visitors to the site. Scriptwriter Frank Higgins and producer Ted Woodward (of Poppycock, Balderdash, & Pshaw) provided the authors with different perspectives and an enlivened view of the site. Academic administrators at the University of Missouri–Kansas City (UMKC) enabled the authors to initiate scholarly joint ventures when they granted Louis community leave to work with Ann. Ann gained insights from a tour of mills in the United Kingdom, a trip supported by the Watkins Mill Association.

Along the way, coworkers and students provided the sustaining skepticism and enthusiasm to spur us on. Among the former were John Stack, Richard McKinzie, and Linna Place (UMKC), Booker Rucker, Larry Grantham, and Mike Beckett (at Department of Natural Resources), Laurence Gross (then at the Museum of American Textile History), and Norman Crockett (University of Oklahoma). Among the latter were Sharon Bagg, Patrick Clark, Clifford Henderson, Lee Dye, Janet Russell, Gail Green, Debbie Welch, Fionna Beattie, and Chad Means. Word processing done by Annetta Potts and Rosalie Donovan often inspired us to take another look at our research.

As it was transformed into a book, anonymous referees sparked us to improve our manuscript. The professionals at Truman State University Press—Paula Presley, Nancy Rediger, and Judith Sharp—further refined our interpretation with the assistance of Cheryl Gracey. The text is deeply enriched because three talented and patient folk, Mike Beckett, David Greene, and Fionna Beattie, combined their artistic eyes and technological know-how to add the images. We are obliged to the patronage of the Watkins Mill Association for underwriting these photographs.

Lastly, we are most appreciative of all the visitors to Watkins Mill who asked questions that sent us back to the sources to discover something we had never thought of before. It is because they were so curious that we have had so much fun!

Louis W. Potts
Ann M. Sligar
June 2004

1

Driving on in the Same Old Way

Walthall "Waltus" Lockett Watkins (1806–84) was one of many restless and energetic Americans who, in the early nineteenth century, forsook family ties in the settled regions of the United States to join a great migration westward into the newly acquired Louisiana Territory. In so doing, they were fulfilling Thomas Jefferson's vision of an "empire for liberty" and advancing the political heritage of a fledgling American republic.[1]

Watkins's destination in 1830 was Clay County, Missouri, at the time the westward margin of the nation. There he faced literal and figurative crossroads, not only which trail to take across the land, but more importantly, whether to pursue farming or manufacturing. Just as Jefferson in the 1790s had experimented with a small nail factory at Monticello, and then as president supported manufacturing as a means of sustaining independence, so would Watkins, as entrepreneur, strive to combine agricultural and mechanical ventures.[2]

Early Life of Waltus Watkins

Born near Versailles, Kentucky, on 30 October 1806, Walthall Lockett Watkins was the third child and eldest son born to Benjamin and Jane (Minter) Watkins, who had migrated from Chesterfield County, Virginia, to the bluegrass country of Woodford County.[3] The record of Waltus's first twenty-five years of life is sparse. Benjamin would not have his son be a mere farmer, but instead arranged for him to receive a rudimentary education at a country day school, followed by a series of apprenticeships. Through farm work, Waltus became familiar with the tools, techniques, peoples, and animals that comprised the art of an agricultural regime, and he apparently acquired skills as a surveyor while laboring under the tutelage of his father. At some point, he was given a ciphering book and measurement charts that his father had used between 1788 and 1793. The book included neat and carefully handwritten columns of calculations, tables for monetary conversions, and systems of dry and wet measures, as well as practical problems and rules for barter.[4]

A family legend claiming descent from "a sterling race of men, with whom matters of principle were supreme to everything else" is key to an understanding of Watkins's character and achievements. In a series of brief autobiographies penned or dictated later in his life, Watkins sought to affirm his rights and responsibilities as a member of the American gentry. Although his contemporaries were prone to depict self-made men who had risen to eminence, Waltus chose to emphasize his family lineage before going on to describe his own career.

The story was most probably passed orally from generation to generation; by Waltus Watkins's era, it could be found in print among the publications the Watkins family had collected. Waltus was the sixth generation tracing back (on his mother's side) to Barthelemy Dupuy, who was characterized as "a distinguished representative of that fearless and true sect of Frenchmen," the Huguenots. Though a legal subject of French monarchs and a valiant military commander, Dupuy and his wife, Susanna LaVillan, were driven from their native land by Catholic persecution after the revocation of the Edict of Nantes in 1685. The foundational legend, as reported by Waltus, pivots on a breathtaking scene where Dupuy, with pistol and sword, fights his way to a ship to escape. Susanna, "a countess of much influence and distinction," might have died in the struggle when pursuers fired at the couple, but, providentially, a "ball had lodged in a small psalm book that she had concealed in her bosom. O, the mercy of God to those who put their trust in him."

The melodramatic lessons learned by the Watkins clan were both moral (Stand up for your beliefs no matter what) and social (Assert your right to prominence and success). In 1870 Waltus remarked, "My father was... too young to take part in the War for Independence, but my relatives one generation older, were all rebels against their government and fought for Independence; and most of their descendants still resist tyranny, come from what quarter it may." Waltus Watkins apparently ran true to form. Family legends, Waltus's autobiography, and his obituaries all characterize him as a man of high principle, determined enterprise, and a community pillar.[a]

[a]Family lore claims that the Welsh origins of the Watkins surname translate to "valiant leader." The name in Virginia has been traced to an indentured servant, Henry Watkins, an immigrant who arrived sometime between 1656 and 1664. Means, "Edited Watkins Family Letters," 2–4; *History of Clay and Platte Counties,* 505; "Old Men of Clay County." Dupuy's escape, minus Victorian melodrama, is recounted in Baird, *Huguenot Emigration,* 109–10. The Dupuy family tree was printed in 1860, and a copy was acquired by Waltus Watkins.

By age 18, Waltus was apprenticed to maternal uncles, who introduced him to manufacturing at many levels of the term. His initial training was at the cotton mill of Jeremiah Minter near Bardstown, Kentucky, where he served as apprentice mechanic for seven months. He then moved under the supervision of James Minter, who possessed "a larger establishment and improved machinery." There Waltus learned the trade of mechanician (one who could not only repair machinery but also solve production problems). As a budding artisan, he practiced both physical and mental skills, becoming accustomed to working with his hands, seeking to see how things work, and using mathematics and his mind's eye to solve problems.[5]

A later biographical profile claimed Waltus had worked as a young man in the cotton and woolen industries. Waltus "made himself complete master of his trade" in the Lexington, Kentucky, cotton mill owned by Charles Lewis. Waltus gained further versatility as a mill foreman and through an unusual experience with steam power. Recognized for his mechanical aptitude, he was enlisted by Joseph Bruen and Charles Lewis, who built a prototype of a steam-powered railroad. Nothing further came of this instance of innovation, but Watkins had glimpsed that remarkable engine of change for his generation: steam power.[6] Through these experiences, Watkins became familiar with the mechanics, tradesmen, and entrepreneurs who comprised the manufacturing interest group in Kentucky. Innumerable artisans such as Watkins contributed their skills, which "can be thought of as a mix of different portions of four components: dexterity, judgment, planning, and resourcefulness." Incremental innovations in production techniques were matched by similar improvements in organizational skills.[7]

Following his apprenticeships, which stretched between 1824 and 1828, Waltus was ready to make his mark. Unlike his father and brothers, who would follow Henry Clay (a very distant kinsman) from the National Republican to the Whig parties, Waltus cast his lot with Andrew Jackson in 1828. Waltus would be a lifelong Democrat. This mark of political independence was an indicator of his maturity. In 1828 he also moved northwest to Frankfort, Kentucky, where he was employed in Colonel Dudley's spinning mill. A nephew who visited Watkins in Missouri in 1873 mused that his uncle "must have [been] a gay lark in his young days judging from the playful deviltry that now and then crops out when the humor takes him."[8]

Later in life, Waltus Watkins dutifully reflected on his upbringing: "The early truths taught to him by his parents were never forgotten or abandoned. Industry, integrity, love of truth, honor and punctuality—these were the lessons he imbibed, and their practice made him ever respected and honored by his neighbors." Such codes of behavior gained Watkins prestige in society and signified self-control. The single extant letter Waltus saved from his father encapsulated the admonition: "Self knowledge is more valuable than all the wealth of India.... I do hope that you may take a few moments every day so long as it is your lot to live, to think what you are and what created for."[9]

Death of Waltus's Father

Waltus's responsibilities were interrupted in 1830 by the death of his father. A local newspaper published a brief obituary of Benjamin Watkins with an accompanying acrostic. The former assured the community of Benjamin's salvation gained by his Baptist faith and limned, "He was a good citizen, tender parent, affectionate husband, kind master, and charitable neighbor." The latter concluded, "Redeeming grace and dying love / shall be my theme in realms above." Both the obituary and acrostic are formulaic and reveal nothing personal of Benjamin Watkins.[10]

A sense of impending death had prompted Benjamin to pen his will in 1830. He left only three specific requests: (1) that the residence and farm "shall remain for the purpose of raising maintaining and education" of his six youngest daughters, and that the executors "give to my daughter Margaret [who had been lamed by a childhood accident] a greater portion of my estate than my other children are to receive, on account of the afflicted situation"; (2) that the older children (three sons and three daughters), as they reached lawful age or were married, could draw on proceeds from the farm for $100 cash and "a horse saddle bridle Etc Etc to any amount not exceeding one hundred & twenty dollars"; and (3) that when the estate was finally liquidated—then anticipated at Jane's death—all the children would gain an equal share.[11]

An inventory of his estate, completed within two months, listed seven slaves, sixty hogs, six horses, sixty-four sheep, twenty cattle, and a yoke of oxen. Additional assets were twelve bedsteads, various farm implements, and household furniture. The last included a loom with gear and a girthline loom, the total value of which was $310. Not counting notes of obligation or the farmstead, the estate was estimated at $3,197.08. The distribution of these assets would be determined by the executors: widow Jane Minter Watkins, sons Walthall, Benjamin (hereafter referred to as Ben), and James, and son-in-law John G. Handy. After paying funeral expenses and "other just debts," the executors had at their disposal the residence and farm at Mount Althea (later estimated at 134 acres valued at $35–$50 per acre), seven slaves ($1,535), and two additional pieces of property (a house and lot in Versailles and a share in the property "now held & occupied by Mrs. Margaret Towler") as well as income from the sale of perishable articles.[12]

John Handy. DNR, WMSHS.

The key decisions regarding Benjamin's legacy appear to have fallen to Jane Watkins and John Handy, as Waltus and Ben Watkins had moved away by 1832. The remaining executor, James Watkins (just 21 years old in 1832), stayed on the Woodford County farmstead until 1837 but differed with John Handy's strategy in handling the family estate, particularly on the question of accepting Kentucky bank scrip versus hard money payment for the farmstead. It was Handy's central task to locate a buyer, one Henry Moss (who had witnessed Benjamin's will), and negotiate the terms of the purchase. The transaction was jeopardized in May 1837 with the onset of a financial panic nationwide, and later by a drought that cut the yields of the farm. Moss could muster only $1,500–$2,000 of the purchase price and threatened suit to evade his obligation. The main points of contention were the fact that prices had fallen precipitously (whereas Moss had bid $80 per acre in 1832, by 1837 comparable land was pegged at $35) and the contract's stipulation that John Handy had to accept Kentucky money. Handy persuaded his kinfolk that these terms were the best to be gained amidst bad conditions. In October, he advanced Moss sufficient money to cover the initial down payment on Mount Althea (subsequently renamed Moss Hill) and took up lodging there while the Watkins women readied themselves for the trek to western Missouri. Moss pledged to pay three annual installments at $35 per acre to complete the sale.[13]

Jane Minter Watkins. DNR, WMSHS.

Moving West, Getting Ahead

Within months of his father's death, Waltus, in counsel with his mother and his siblings, determined to venture to Missouri. He and brother Ben visited Liberty in the fall of 1830. Returning to Kentucky, he sought to raise capital from notes held by his father, pleaded for additional funds from his siblings, and viewed a cotton spinning mill run by a relative near Nashville. Finally he set off to Pittsburgh and Louisville, where he purchased machinery for his Missouri venture.[14]

In his move west, Waltus Watkins joined a torrent of pioneers. The path he followed from Kentucky to Clay County was a popular one. Missouri's natural attractions included its extensive river system, fertile bottom lands, abundant timbers, temperate climate, and moderate growing season—in other words, a landscape similar to the one Kentuckians had left behind. From 1815 until 1860, roughly sixty thousand people per year migrated to the trans-Appalachian West. In the economically distressed 1830s, an estimated 5.8 percent of the population of Kentucky joined Watkins in moving to the West; one contemporaneous observer of this "Great Migration" noted, "Kentucky and Tennessee were breaking up and moving to the 'Far West.'" Between 1820 and 1860, the population increase in Missouri was seven times that of the nation.[15]

For many pioneers, western land prices provided an incentive to move: $1.25–$2.00 per acre in contrast to the $30–$40 per acre in the eastern states. Men from the southern Piedmont sold their farmsteads at a relatively high price, thereby bringing new wealth with them when they migrated to the Missouri frontier. In a culture accustomed to large families, extensive land holdings symbolized not only independent livelihood but also security in one's old age.

A combination of forces propelled Watkins westward. Like many others, he recognized that the West could provide economic advancement plus upward mobility for himself and collective security for his kin. Equally attractive were the prospects for commercial capitalism, as there would be little competition for the milling skills and services Watkins could provide to customers on the Missouri frontier.

While the vast majority of those on the move west were in quest of the newly opened agricultural region, Watkins represented the initial wave of industrialism. Though still honored as the keystone of the republic's economy and political system, the yeoman farmer was being challenged. In town and city, artisans and merchants were caught up in a market revolution. The imperative was to increase the volume and decrease the price of production of consumer goods; mechanization took hold as the means of accomplishing these goals. As one analyst has put it, "The greater profits now realized through standardizing and mechanizing production with low-skilled labor were, in part, returned to customers in still cheaper goods that further undercut the serving crafts." (During his apprenticeships in Kentucky, Watkins had glimpsed how his uncles had fallen victim to the market cycle.) The economic transformations of the 1820s and 1830s unfolded amidst untold numbers of small innovations as well as major inventions, and comprised the early stages of the American Industrial Revolution. In a land where there had been a labor shortage since colonial days, schemes to remedy underemployment or mechanisms devised to save labor were particularly applauded. The very concept of "industry" was changing. Whereas the word had once been used to describe cleverness, dexterity, or ingenuity, in Watkins's lifetime its meaning incorporated a systematic approach to work wherein managers or workers who were dynamic, efficient, and productive were often labeled "industrious."[16]

The irresistible sweep of western migration pulled Americans through gaps in the Appalachian chain and dispersed them through the valleys of the Ohio, Mississippi, Missouri, and lesser interior rivers. As the frontier zone pushed westward, equally radical changes affected communities; the simpler values and relationships of a rural world were challenged by the complexities of urbanism and industrialism. A transportation revolution heralded by steamboats and railways provided the means for swift changes. Mechanized industry, monetary credit, and a price system for goods and services comprised the emerging market economy. "Americans sensed that they were living in a new era. Some welcomed it but many feared the disruption of traditional social relationships."[17]

In the forefront of these changes were expectant and opportunistic capitalists such as Watkins, who sought to grasp the promise of wealth as a return on their aggressive enterprises. Of America's capitalist development between 1830 and 1860, analyst Anne C. Rose has observed, "Not everyone was an eager entrepreneur.... Profit-seeking sorely challenged the attention to spiritual growth at the heart of evangelical Protestantism, and competition threatened to make the vision of a harmonious polity that had inspired the founders' classical republicanism obsolete.... In all cases capitalist values mediated between Americans' hopes and fears and between their ambitions for change and attachment to tradition." Waltus Watkins, like innumerable members of a new generation of democratic Americans, could be described as enterprising, improvement-conscious, and energetic. He scrambled to get ahead.[18]

Clay County, Missouri

Missouri, from acceptance into the Union in 1821 until the conclusion of the Civil War in 1865, played a unique role in the nation's history. It lay at the intersection of two distinct geographical axes. The first was the east-west axis along which the forces of the nation's westward movement traveled. Politicians from the state consistently supported expansion of the Union. Clay County would be particularly proud of Alexander Doniphan's exploits with his company of 109 men in the War with Mexico (1846–48). Senator Thomas Hart Benton was an enthusiastic orator for the push west. National expansion placed Missouri in the critical role of middleman, receiving commodities from the West and sending them to eastern markets, a position that only diminished with the completion of the transcontinental railroad after the Civil War and Chicago's triumph over St. Louis as commercial nexus.

The second axis was north-south. As a border state, antebellum Missouri evinced the entire range of arguments swirling around the issues of slavery and states' rights. The eventual consequence—a civil war within the state and the nation—brought unprecedented agony to Missourians. In this crossroads state, competing cultures collided as slaveholders encountered abolitionists, recent immigrants from western Europe clashed with families who traced their roots to colonial America, and, as personified by Waltus Watkins, industrialism confronted agrarianism.[19]

Watkins's reconnaissance of western Missouri in 1830–31 might not have been hospitable. The winter's snows started in October and were recorded in subsequent county histories as impressively deep (two to three feet with drifts to fourteen feet). The following crop year was largely noted as a failure for corn. Nonetheless, Watkins and other

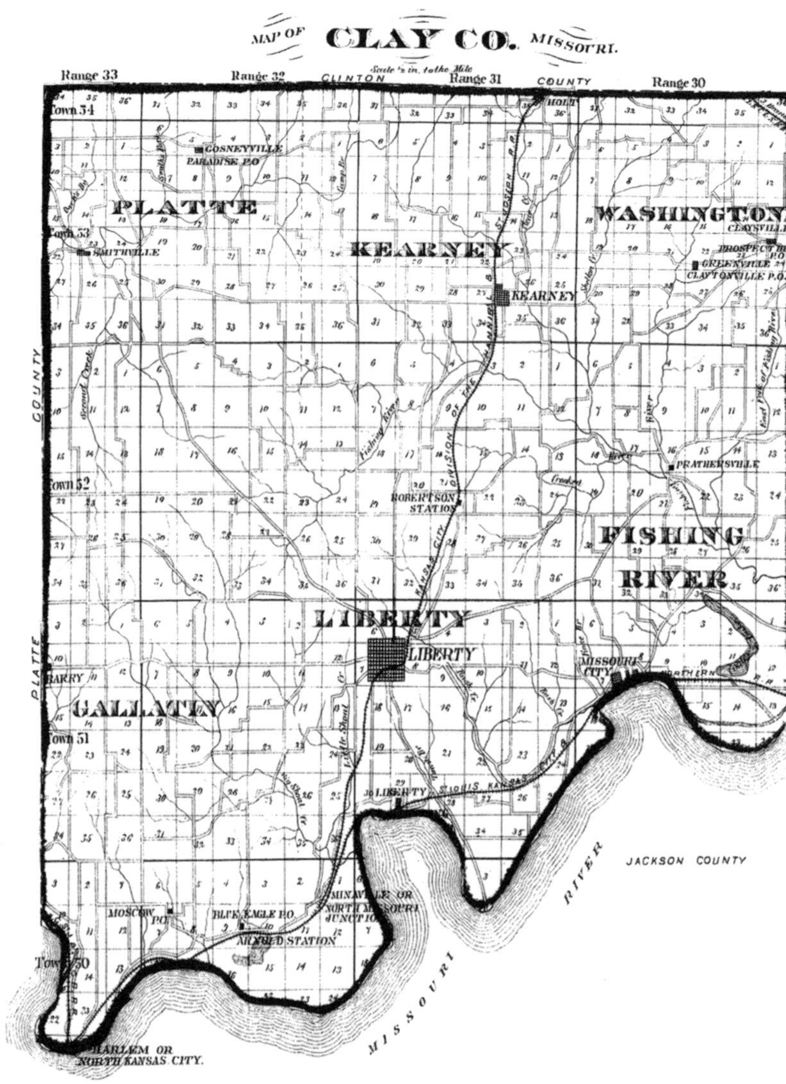

Map of Clay County, 1877. From the *Illustrated Historical Atlas of Clay County, Missouri.*

migrants flocked to Clay County, established in 1822 as the westernmost extent of Missouri. Clay County was largely shaped by transportation routes and by the values of its initial settlers. One of Watkins's cohorts, Joseph Thorp, described these pioneers as "mostly Kentuckians and Virginians—sons of the old pioneers of the canebreaks of Kentucky and the mountains of Virginia. They were generally poor, but possessed of intelligence, industry and integrity, and were fearless and determined, relying on their own manhood for success." The county's name indicated the esteem that Missourians felt for Kentucky Congressman Henry Clay, who had particularly championed Missourians' cause for unrestricted statehood in the controversial years 1819–21. Naming their county seat Liberty echoed the settlers' reverence for their revolutionary forebears and their current political stance.[20]

A woman who came to Clay County in 1821 believed the county's pioneers were no different than other American migrants in terms of character, but rather were set apart by two key facts. First, the soldiers quartered at Fort Leavenworth, Kansas, provided peace; she had "no fear whatever of the Indians." (Indeed, in the extant Watkins family letters, there is no mention of Native Americans.) Second, town life thrived in the county seat at Liberty: "The soil is fertile and the country generally healthy, with a good frontier market for all kinds of produce.... It was so new and everything seemed bright—I doubt if there was in any other town in the United States so many brilliant men and women as Liberty had." This reminiscence was echoed by another old settler who recollected of 1829 that "Liberty was the paradise of western towns, the ultima Thule of western emigrations." Such boosterism gained momentum to the 1860s. Clay Countians, especially after 1846 when editor Robert Hugh Miller began the *Liberty Tribune,* were "not willing to acknowledge any superiority of soil, climate and population to any nation, kindred, or tongue under the sun." Not everyone was so taken with the town. In June 1831, after making the seven-week round-trip to accompany Watkins and his newly purchased machinery, John Handy told Waltus's sister Martha that he considered "Liberty a most dissipated town, with no opportunities for his carpenter trade, and while St. Louis is somewhat better, [he considered] it an unhealthy place to take his family."[21]

Joseph Thorp described Clay County in 1825 as "truly a land of Canaan, for it flowed with milk and honey...abundance of game...rich soil." However, there were some drawbacks to economic development: salt was difficult to obtain, greenhead flies bedeviled farmers, and wagons had difficulty finding reliable supply routes. The onset of trade with Fort Leavenworth (established in 1827) and the acquisition of the Platte Purchase farmlands just to the west in the decade of the 1830s both proved godsends to the county. The former added an exotic blend of fur traders, who often alighted with their goods at Shubael Allen's steamboat landing; the latter required supplies of all sorts. Richfield (later called Missouri City) became a key port for river traffic, and the merchants and craftsmen grouped at Liberty served the road that led between there and Fort Leavenworth. The county's population, despite formation of other counties from its territory, mounted from 5,338 in 1830 to 9,426 in 1848, an increase of 44 percent in eighteen years.[22]

Gazetteers of the state of Missouri, aimed at wooing settlers, provide clues as to inducements and risks. One, in 1823, noted of Clay's prairies, "Perhaps before the expiration at ten years, instead of being bleak and desolate, they may have been converted into immense grazing fields, covered with herds of cattle." The closest, "smaller ones...are by far the most desirable and pleasant situations for settlement." A similar publication described Clay in 1837 as "now one of the best settled tracts of country in Missouri or elsewhere. The high cultivation of numerous and large farms, the substantial buildings and the tasteful arrangements about the domiciles of the old settlers." This development was traceable to two sources: "Great wealth was carried to the country, and more had been acquired by the enterprise and industry of the inhabitants." The vision of the 1830s inhabitants was to model Liberty on Lexington, "the Athens of Kentucky."[23]

The main lure for Waltus Watkins was that Clay County in its first decade was deemed "fine cotton country." Like other entrepreneurs, he aimed to process the products of the market economy that arose quickly there as in other "Little Dixie" counties along the Missouri River. Grain mills, tobacco stemmeries, rope walks, and wool and cotton carding or spinning establishments were scattered among the rude log cabins in the area or congregated with the grocers, saddlers, and wheelwrights at the county seat. "Freight was high but money was plenty, and everybody thought there was no such place as Clay county" was the impression of the early 1830s. "The thought of ever being in want of a market for the surplus productions of the county never once entered into any of our minds."[24] Yet through the 1830s and into the next decade, according to a later county history, "the want of such mechanics is seriously felt by merchants, families, and farmers" of Clay. The U.S. census of 1840 clearly indicates the allocation of the county's workforce: Agricultural workers numbered 2,900, compared to 57 engaged in commerce and 192 in manufactures and trade. Only $22,745 was invested in manufacturing, none in woolen or cotton mills.[25]

Watkins's Time in Liberty

Watkins's years in Liberty (1831–39) marked his efforts to establish himself as an entrepreneur, a family man, and a civic presence. An uncle in Kentucky, Anthony Minter, noted the challenges Watkins faced on the frontier. If local cotton growers failed to produce an ample crop, he would have to rely on the St. Louis or Nashville markets and thus pay additional transportation costs. In 1832, the uncle's prognosis was "When you get settled and have good credit, you can go to the bank and buy [cotton] in advance. But as things now stand, you share the fate of all beginners with small capital. In two or three years, with industry, economy, and common luck your situation should be much better." In the fall of 1831, Watkins started ox-powered spinning operations, with equipment transported from Pittsburgh and Louisville, the mechanical talents of one Anthony Harsell to help him get started, and assistance from occasional rented slave labor.[26]

The young entrepreneur persevered and adapted. As a later profile sketched his endeavors at Liberty: "Here he built a small cotton factory, but the farmers raised but little cotton, as the climate was not adapted to its growth, and he abandoned the enterprise for want of material. He then converted his cotton factory into a wool carding establishment, which though more prosperous, was not sufficient to satisfy his ambition." To augment his eighty-four cotton spindles (the first such machinery in western Missouri), Watkins also ventured capital and know-how to set up a tannery and barkmill in Liberty, perhaps in partnership with one C. Lockwood. The aim of these enterprises was to convert the region's raw materials into hides or yarn for subsequent processing by consumers in the area. In June 1831, he purchased 1.5 acres west of the courthouse square on the north side of Kansas Street from John Peters. There he established his residence.[27]

Mary Ann Holloway Watkins

As Watkins established his cotton mill in Liberty, he received more advice from Anthony Minter, a bachelor: "Try to marry as soon as you can make a good bargain. I don't mean as to property, but as to profitable good sense, someone well disposed toward contentment in her troubles—as none without—have"; and "Remember also that all should prepare for the life to come. It is the only sound policy for man, in his duty to God and man, and the only means of happiness." In the winter of 1833–34, Watkins returned once again to Kentucky where, according to family legend, a sister teaching in Jessamine County, probably Mary, introduced him to one of her students. This was Mary Ann Holloway (1817–96), ten years younger than Waltus and daughter of Spencer and Catherine (Reed) Holloway. A courtship of six weeks led to their marriage on 4 March 1834. Watkins later recollected that his wedding day had been "a beautiful, pleasant day and clean." Soon the newlyweds were on the trail back to the Missouri frontier.[28]

Waltus's view of marriage may be inferred from the prescription he offered when daughter Martha Ann married in 1874: "I have long thought that [the] sure way of making the married life hap [*sic*] is to live up to your duty as laid down in that Book of Books which says husbands love your wives & wives obey your husbands & let the wife see that she reverence the husband and be in subjection to the husband. I have seen many try to take near cuts on this but it wont do all had better obey the divine injunction."[29] It is clear from extant family records that Mary Ann's role as wife and mother was both valued and multifaceted. Although Waltus may have been profiled as a self-made man (in keeping with the common middle-class view of "the independent man as the instrument of progress"), this was partially fictional. For him, as for many of his generation, "success in reality was a family enterprise." A remembrance of Waltus and Mary Ann emphasized that they "were well mated; his business was out of doors, hers in the house. Here her will was law." Another confided that Waltus's success could be traced "largely to the energy and strong character of his wife." Waltus recognized Mary Ann's contributions. In 1864, he

Waltus L. Watkins, c. 1860.
DNR, WMSHS.

Mary Ann Holloway Watkins.
DNR, WMSHS.

acknowledged "your Ma has enough to do to keep 3 women at work."[30]

The children particularly showed affection to their mother and were in awe of her stamina. Martha reminded her father that his wife "dont complain. She is the sweetest most self sacrificing soul in the world." Mary Ann epitomized the piety and domesticity of the Victorian lady. She was skilled at dressmaking and embroidery; her daughters particularly cherished their mother's expert needlework in their lace dresses. She made fine embroidered scarves for furniture in the family's formal parlor. Upon occasion she wrote poetry, and she was a solid pillar of her church. Her obituary marked her as "one of God's noble people."[31]

FAMILY MIGRATION

The Watkins clan was ambitious, reflected in a refrain shared in Holloway and Watkins correspondence: "We are all moving in the same old way," or "We are driving on in the same old way." In the decades of the 1830s and 1840s, the extended family, in a pattern of chain migration, followed Waltus to western Missouri. By 1850, most of Waltus's close relations were settled in Clay and Platte counties, many within short distances of each other. These included the matriarch, Jane Minter Watkins, and eight of her children, together with their spouses and offspring. Waltus and Mary Ann not only spearheaded the migration but also gave essential assistance as the various family units arrived. They provided his mother with a base, first in his crowded cabin in 1839 and later in her own room in the family's permanent house, called "the Big House." From there, Jane and her daughter Rebecca traveled on circuit through Clay, Clinton, and Platte counties, making extensive stops between 1839 and 1859 with her children and other kin—Minters, Carters, Lukes, Gills, and Blythes. Waltus's sister Margaret would refer to herself and her sisters as his "corn cracker relations." Sisters in Kentucky lamented to Missouri kin, "[We] never knew until we were separated how happy we were when together...in a family circle at home."[32]

The decision to move did not necessarily come easily. In 1838, Margaret, about to join the chain migration, asked for "a long letter" from kin in Clay County "and tell me how people live in Mo., so that I may prepare my system for a change by the time." Another sister, Caroline, preferred free-state Ohio to slave-state Missouri: "I have always thought it wrong to hold slaves and I believe we would do better without them than with them. And I expect in less than ten years the Americans will be sorry they ever saw one." (In spite of these sentiments, within ten years Caroline and family had moved to Platte County, Missouri.) A third sister, Elizabeth, initially remained behind. Then in 1846, torrential rains devastated a sawmill on the farm of her husband, Isaac Carter; a year later Elizabeth confessed to Waltus of "Mr. Carter's misfortunes embarrassments and also of his being quite dissipated." Handy intervened to rescue Elizabeth and her children, keeping them in his home and selling her slaves. She made her way to Missouri, where Waltus purchased an $800 farm for her using funds secured by Handy. She evaded her estranged husband when he ventured to Clay County but later remarried him and lived northwest of Waltus at a farm she called "Woodland Cottage."[33]

Waltus's oldest sister, Mary Handy, remained anchored in the Bluegrass region west of Lexington, Kentucky, where John Handy speculated in farmsteads. Mary reported to Waltus in 1843 that her husband "sometimes gets the Mo fever very bad, but I believe it gradually wears off again when he undertakes a big job of work.... Mr. Handy says if people here rob him of much more money he will never be able to buy a farm in Ky and Mo will be the highth of his ambitions." John Handy navigated through financial perils to prosper as he moved his family frequently between Jessamine and Woodford Counties to ever better homesteads. When Waltus's son George first visited John Handy in 1856, he wrote home that his uncle was "a man of vast experience and he seems to know about almost every body [*sic*] in Kentucky."[34]

The Watkinses traveled the Missouri-Kentucky route periodically. Mary Ann returned home from western Missouri for the births of her first three children, in 1835, 1837, and 1839; each time, she headed

In their travels, the Watkinses took advantage of what historians call the Transportation Revolution. In 1840, John Handy recounted that it took him eight days to ride horseback across Missouri from Watkins's cabin to St. Louis. It must have been a familiar route, for he remarked on lodging or stopping for breakfast with people Watkins knew. "[At Franklin I] rode up to the house that we put at when we first traveled to Mo but as soon as I recognized it as being the same I rode off to another Inn; recollecting that we had to wait until eleven oclock for supper then to sleep on the floor."

In 1847, Waltus's sister traveled by steamboat from Louisville to St. Louis and then to Richfield, the latter leg taking nearly six days because of snags and low water in the Missouri River. "I enjoyed my steamboat trip quite well all things considered," Elizabeth related. The Missouri boat "was very much crowded there were about 70 ladys besides men negroes and children." Music and dancing filled each evening; periodic mishaps kept the voyage from becoming dull. Elizabeth reported that a man had fallen overboard and drowned, leaving an especially bereaved widow, for "it was said that he had a large amount of money belted around him." At St. Louis, where the boats "were strung up and down the river for miles as close as they could stand," her boat lost its wheelhousing in a collision. (During another trip Waltus's mother and sister survived an explosion on their steamboat.)

By 1875, when Waltus and Mary Ann came back home from Kentucky, they could rely on railroads. It was an overnight trip from Louisville to St. Louis and another overnight trip from St. Louis to Lawson, where John would fetch them in the family carriage.[a]

[a]John Handy to Waltus Watkins, 4 November 1840; Elizabeth Carter to Mary Handy, 25 November 1847, Waltus Watkins to John Watkins, 12 March 1875, WMSHS.

back to Missouri on horseback a week later. Watkins and John Handy made the trip repeatedly for business reasons during the 1840s and 1850s. Jane Minter Watkins, Mary Ann Watkins, and later Mary Ann's daughters particularly relished trips to the Handy kin in Kentucky, as the families shared bourgeois tastes and habits. Mattie Watkins reported home, "A great many persons said the Handy delegation of women was the handsomest on the ground.... Now dont you imagine we all felt very elated at such a compliment coming from so many?"[35]

Money Troubles

The matter of credit was a critical factor as Watkins began his career as an entrepreneur. In the early nineteenth century, credit was founded on direct personal knowledge of the good character and sound business habits of the applicant and might be gained from a network of kin, friends, business acquaintances, or customers. As emigrants like Watkins distanced themselves from their home bases, efforts to obtain credit increasingly involved unknown and impersonal factors.[36]

Waltus's initial main source of credit was the legacy of his father, a source not without its problems. Settlement of Benjamin Watkins's estate dragged on and on. In 1839, John Handy estimated the proceeds at $7,500 and attempted to bring things to closure. In 1840, soon after a visit to Waltus, Handy proposed buying out any heir but noted, "My proposition to buy out the heirs does not include the property and money in the possession of Mrs. [Jane] Watkins." He was stymied by depreciating script in Kentucky, by the reluctance of banks to subscribe notes, and by a lack of secure messengers to carry negotiable paper to his Missouri relatives. Finally, in May 1841, he sent a Philadelphia check for $1,892.40 to be distributed to four heirs, at the same time reaffirming his buyout offer. Ben, Elizabeth, and Margaret settled for $1,520 apiece; James and Waltus, in separate prior agreements with Handy, had borrowed against their shares to finance their Missouri endeavors. After grappling with the vexing problem for more than fifteen years, Handy reflected, "It is a matter of sorrow and regret with me that I should have so much trouble and labour in settling your fathers estate and cannot Render satisfaction to all the heirs. But I done the best that I could and more than they ought to have expected."[37]

Watkins traced his friendship with, and dependence on, John Handy back to 1824–25; Handy was for many years the central player in his brother-in-law's credit situation. In an era short on cash and prone to unpredictable credit—as a Clay County contemporary reported, "Money is the dearest [here] I ever saw it. It is not to be borrowed at no per sent for it aint here to lend"—Handy's support was critical to Watkins's entrepreneurial drive. As noted, Watkins secured a loan from his brother-in-law based on the pending settlement of his father's estate; by 1 March 1844, Waltkins was in debt to Handy for $3,434. Periodically he sent cash back to Kentucky for which Handy added 10 percent discount, against the 6 percent annual interest on the note. By 1 March 1850, the balance was $3,030.34. This indebtedness was part of numerous loans extended by Handy, who occasionally prodded his kin to make payment but never wrote forcefully for collections. In 1845, Handy, although harassed by his own creditors, confided to Watkins, "I do not wish you to make any sacrifice but as soon as you can pay without sacrifice I wish you to do it." (This pattern continued even after Handy's death in 1867. At the end of 1876, Mary wrote, "Now Brother, let us say a few words about our money matters...times are very hard in MO, & I don't want you to pay more interest than is legal here." She proposed that for the $2,000 loan she had granted him in 1872, he pay up the 10 percent interest and thereafter renew it at 8 percent. A year later she rebuffed his request for additional funds.)[38]

Other than Watkins's indebtedness to his relatives, his early liabilities cannot be ascertained. In 1846, Handy wrote Watkins, "Benj will soon be able to get out of Debt with good luck—& I do Expect he will

Money troubles caused a number of strains in Watkins family relationships. For one thing, Waltus's alliance with John Handy did not sit well with brothers James and Ben. In 1832, Ben lectured Waltus to remember "the pious admonition that our dying father gave to us upon his death bed, which was, that we should all live in brotherly attachment, love and esteem for each other." Ben vied with Waltus for control of capital distributed from their father's estate, for example trying to persuade his siblings to invest "in Woodford [County] at mothers and other places," rather than in Missouri lands or Waltus's enterprises. In 1839, the Handys sought to shift $1,000–$1,200 from Ben's control to that of Waltus: "We are desirous of putting the money in your hands if you want it and can use it profitably." Still, John Handy expressed faith in Ben, claiming in 1840, "I would not refuse to lend him any amount of money.... He is a safe calculating man in my opinion." That opinion had altered by 1847, when Waltus journeyed to St. Louis, where Ben had established himself, to escort their sister Elizabeth.

What split the brothers was a sum of $500 that John Handy had sent to Ben to be passed along to Margaret Watkins Byram. When Ben, for more than two years, did not transmit the money, John Handy caught the brunt of Byram's attack. Regarding the trip to St. Louis, Ben wrote to Mary Handy that he had "an interview with Waltus concerning our difficulty" in the presence of Elizabeth; Elizabeth, for her part, reported that Waltus and Ben "talked over their hard feelings, but did not make it up." Waltus refused to enter Ben's house. Accusations flew. In Ben's version, his older brother "was in the fault and where the rub laid." Waltus refused the option "to let the matter drop forever." Rather than submit their contention to a magistrate, Ben predicted "I calculate on nothing else but his crawfishing out of it." If the matter were not settled, Ben threatened, "I am determined that he shall not trifle with, and insult my feelings any longer without being made to sweat for it." It would not be until 1850 that Ben and Waltus patched up their hard feelings.[a]

[a]Benjamin Watkins to Waltus L. Watkins, 18 March 1832, John Handy to Waltus L. Watkins, 2 November 1840, WMSHS; Elizabeth Ann Watkins Carter to Jane Minter Watkins and Mary Watkins Handy, 25 November 1847, in Means, "Edited Watkins Family Letters," 108–12; Benjamin Watkins to Waltus L. Watkins, 6 January 1850, Benjamin Watkins to Mary W. Handy, (1846?), WMSHS.

Beat you in that matter." In Watkins's extant records there is only one reference to explicit anxiety over credit. A biographical profile, done when he was more than 70 years old, related that in 1860–61 when he equipped his woolen mill, "the war coming on, those who were indebted to him were unable to pay, and he found himself straitened for means. His long life of probity enabled him to borrow money and his energy and industry again came to his rescue, and he became very prosperous in business."[39]

A serious financial setback befell Watkins in 1839. He had negotiated sale of his Kansas Street premises in Liberty to a T. Haile via a deed of trust. When the mill building burnt before the sale was completed, Watkins "lost the fruits of his labor." He returned the purchaser's funds, then resold his Liberty house and mill site. Though Watkins family lore remembered this near-tragedy as just another obstacle to be overcome, it brought 33-year-old Waltus Watkins to a turning point in his life. His career in textiles, despite his training and support from family, had been a failure. The region of Clay County could not support his endeavors in cotton and woolen processing during the depression-wracked 1830s. His toils as an early industrialist had seemingly ended in the fire that burned him out of Liberty.[40]

A later assessment marked Watkins as "a man of great energy, [whose] life had been a busy one—whatever his hands found to do, which he thought would be beneficial and honorable to himself or the people, he done with all his might." It was in that spirit that he now, assuming the role of patriarch of the extended Watkins family, would relocate to lands in the northeast sector of Clay County and seek to reestablish himself, this time as an entrepreneur in agriculture.[41]

Notes for Chapter 1

[1]Jefferson, Letter to Madison, 27 April 1809, Library of Congress, http://www.loc.gov/exhibits/jefferson/149.html (accessed March 2003).

[2]"The Debate over Manufactures in the Early Republic, 1785–1820," in M. R. Smith and G. Clancey, *Major Problems,* 103–41; Yarbrough, *American Virtues,* 77.

[3]Watkins's christened name is traceable to his paternal grandmother, Mary Walthall Lockett. It was probably in early adulthood that he began experimenting with variations, first Walthal or Walter and then fixing on Waltus. For years, his siblings addressed letters to him as Walter or Walthall. See Roney, "Old Watkins Mill," 5.

Benjamin Watkins was born in 1775, Jane (daughter of the Reverend Joseph Minter and Jane Trabue) in 1781. They were married 21 April 1801, in Chesterfield County, Virginia. The couple was following kin in migrating to Kentucky. The farmstead, designated Mount Althea (Althea being the Latin name for the Rose of Sharon plant, a specimen of which was later transported to Missouri by members of the family), was first owned by Benjamin's uncle John Watkins, who sold it to another uncle, Henry, who in turn sold it to Benjamin. Means, "Edited Watkins Family Letters," 2–3; "Waltus L. Watkins of Northeast Clay County, Mo.," WMA Research Files, WMSHS.

[4]McGraw, "'So Much Depends,'" 95; Benjamin Watkins cyphering book, WMSHS.

[5]*U.S. Biographical Dictionary.*

[6]*History of Clay and Platte Counties,* 505; *U.S. Biographical Dictionary;* Addendum to *U.S. Biographical Dictionary,* in Roney, "Old Watkins Mill," vi, 3, 4; Watkins did not record any impressions of a steam-powered wool carding factory established in Lexington in 1829. "Old Men of Clay County."

[7]Licht, *Industrializing America,* 40–41; Gordon and Malone, *Texture of Industry,* 38–39.

[8]Caldwell, "Watkins Mill," 365; "Old Men of Clay County"; George Handy to Margaret Blythe, 11 March 1873, WMSHS.

[9]Benjamin Watkins to Walthall Watkins, 4 June 1825, WMSHS.

[10]Benjamin Watkins obituary, 23 June 1830, WMSHS.

[11]Inventory of Estate of Benjamin Watkins, 15 August 1830, WMSHS; Benjamin Watkins will, attached to letter of John Handy to Waltus L. Watkins, 23 January 1847, WMSHS.

[12]Means, "Edited Watkins Family Letters," 42.

[13]James M. Watkins to Waltus L. Watkins, 8 August 1837, WMSHS; Means, "Edited Watkins Family Letters," 41; Roney, "Old Watkins Mill", i, 30–34.

[14]Oberholtz, "Life Span" and "Early Pioneering."

[15]DAR, *Centennial Souvenir,* 33; Barney, *Passage of the Republic,* 14–30. Among those "old men of Clay County" (60 or older) who responded to a call in 1870 from the *Liberty Tribune,*

67 of 143 reported Kentucky as their birth state. Twelve came from Fayette County, seven from Clark, and six each from Woodford and Bourbon. Fellow Kentuckians David Rice Atchison and Alexander Doniphan, who landed in Clay County about the same time as Watkins, would emerge in the next three decades as the major political leaders of western Missouri, the former as a Democrat and the latter as a Whig. See Potts, "Recollections."

[16]Barney, *Passage of the Republic,* 31–39.

[17]Henretta et al., *America's History,* 1:296.

[18]Rose, *Voices of the Marketplace,* 61. See also Sellers, *Market Revolution,* 152; Wood, "The Enemy is Us."

[19]Stiles, *Jesse James.* Chapter 1 outlines the cultural clashes in Clay County experienced by the Reverend Robert James, father of Jesse and Frank James, who was also Watkins's minister and a fellow migrant from Kentucky.

[20]Roney, *From Entry Fee,* 7; Thorp, *Early Days,* 35, 59; *History of Clay and Platte Counties,* 119, 121–22, 147; McMillen, *Paris, Tightwad and Peculiar,* 50. Stuart Voss notes in "Town Growth" (62) that Clay County was considerably more aristocratic, more Whiggish, and more hospitable to Mormons in the 1830s than its near neighbor, Jackson County.

[21]Petty, "'Some Punkins,'" installments 4, 12, 13, and 16; Robert Miller, quoted in Petty, "Some Punkins," installment 1; John Handy, quoted in Roney, "Early Pioneering."

[22]Thorp, *Early Days*, 35, 59; *History of Clay and Platte Counties,* 119, 147.

[23]*History of Clay and Platte Counties,* 109, 131, 284.

[24]Petty, "Some Punkins," installment 16; *History of Clay and Platte Counties,* 121.

[25]Woodson, *History of Clay County,* 154; U.S. Census Office, *Statistics of Sixth Census,* 356–57, 416.

[26]Roney, "Advice"; *U.S. Biographical Dictionary;* Anthony Minter to Walthall L. Watkins, 28 February 1832, WMSHS; *Liberty Tribune,* 16 February 1883.

[27]Modern theories of entrepreneurship describe entrepreneurs as people who are partial to innovations, to forging new combinations from available tools and techniques, or to opening new markets. Analyst David McClelland (quoted in Pusateri, *History of American Business,* 12) has claimed entrepreneurs have "a desire to do well, not so much for the sake of social recognition or prestige, but for the sake of an inner feeling of personal accomplishment." Watkins's traits were compatible with those McClelland has identified as distinctive to entrepreneurs and defined as "decision making under uncertainty." *U.S. Biographical Dictionary;* "Waltus L. Watkins Land Transactions," Research Files, WMSHS.

[28]Anthony Minter to Waltus Watkins, 28 February 1832, WMSHS; Waltus L. Watkins to Mary Ann Watkins, 4 March 1881, WMSHS.

[29]Waltus L. Watkins to Mrs. M. D. Scruggs, 18 December 1874, WMSHS. Mary Ann was not one to write or to talk much about herself and hardly added to the voluminous correspondence her husband produced.

[30]Rose, *Voices of the Marketplace,* 77; Mary Ann Watkins obituaries by G. L. Black and T. P. Stafford, Mount Vernon minute book, c. 9 September 1896, WMSHS; Waltus L. Watkins to Kate and Mattie Watkins, 3 June 1864, WMSHS.

[31]Martha Scruggs to Waltus L. Watkins, 22 November 1880, WMSHS; Stafford, obituary of Mary Ann Watkins, WMSHS.

[32]Margaret T. Watkins to [?], c. 1837, WMSHS; Margaret T. Watkins and E. W. Carter to James M. Watkins, 26 January 1838, in Means, "Edited Watkins Family Letters," 34–37. Stuart Bruchey, in *Enterprise* (p. 195), focuses on "getting on" as the slogan by which entrepreneurial success was measured in antebellum America.

[33]Mary Watkins Handy and John Handy to Waltus Watkins, 13 March 1843; George Watkins to Waltus L. Watkins, 1 January 1856, WMSHS; Potts, "From Speculator to Gentleman," 46–55. For similar migrations, see O'Brien, *Grassland,* 319–24.

[34]Mary W. Handy [and John Handy] to Waltus L. Watkins, 13 March 1843, in Means, "Edited Watkins Family Letters," 67–70; George Watkins to Waltus L. Watkins, 11 January 1856, WMSHS.

[35]Mattie Watkins to Lizzie Watkins, 7 September 1867, WMSHS. The bourgeois tone of the family life can also be glimpsed in this bit of advice from an older brother: "Be very

careful with our little brothers and sisters in giving them advice and instruction teach them to consider themselves above the vulgar. If you take the right course with them they will follow your counsel more than you think. Kiss my good mama for me she is one of the best mothers in the world." John H. Watkins to [his sisters], 10 May 1867, WMSHS.

[36]Norris, *R. G. Dun & Co.,* xv–xvi. Neither Waltus Watkins nor any of his later firms appear in these extensive credit ratings.

[37]Mary W. and John G. Handy to Waltus L. Watkins, 30 June 1839, 2 November 1840, 20 May 1841, 16 October 1841, and 23 January 1847, WMSHS; Anthony Minter and John Handy to Waltus Watkins, 22 April 1846, WMSHS.

[38]A. Wilson to Samuel Turrentine, 26 April 1840, 10 January 1846, W. L. and J. M. Watkins c/o John G. Handy, 1850, A. Minter and Mary and John Handy to Waltus Watkins, 23 October 1845, Mary Handy to Waltus L. Watkins, 18 December 1876, 21 November 1877, WMSHS.

[39]*U.S. Biographical Dictionary.*

[40]*U.S. Biographical Dictionary.*

[41]Waltus Watkins obituary, c. 24 January 1884, Research Files, WMSHS. That Watkins soon succeeded was noted in an 1870s biographical sketch, which observed that in the 1840s he became "very successful as a farmer, and soon retrieved his losses." *U.S. Biographical Dictionary.*

2

Too Good a Home There to Leave

When he left Liberty, Waltus Watkins did not continue his westward migration, though the opportunity—and perhaps the temptation—was vividly present in the large trains of white-topped wagons rolling westward from nearby Independence on the Oregon-California trail. Nor did he join the nearly nine hundred other Missourians (including Company C from nearby Liberty) who ventured down the Santa Fe Trail in 1846 to join the war against Mexico. Apparently Watkins did contemplate selling his farm around that time, telling his sister Caroline in an 1846 letter, "[It] is so large and so much trouble." Yet he did not leave. He had become rooted in western Missouri, and there he stayed to raise a family, develop extensive landholdings, try his hand at numerous commercial ventures, and establish himself as a prominent citizen of the community.[1]

An Extensive Family

The family that Waltus raised with Mary Ann was an uncommonly large and robust one. Ten months after their marriage, their first child was born; siblings arrived at regular intervals over the next twenty-four years until there were eleven children in all.[2] All but two of the Watkins children lived to adulthood. Life expectancy for Americans in the second half of the nineteenth century was 40–47 years for white males and 43–49 years for white females, yet both Watkins parents and seven of their children lived well into their seventies and beyond. (Three of the offspring died in their eighties and two in their mid-nineties; even with two childhood deaths, the average lifespan of this generation was 61.8 years!)[3]

During most of their marriage, Waltus and Mary Ann provided room and/or board for a large extended family. Three of the children—John Handy, Joe Berry, and Caroline Emma (Carrie)—lived almost their entire lives at Bethany; Waltus's mother spent her last two decades there; and one of his sisters, Rebecca Watkins, resided with the family from 1839 until she married in 1859, at age 38. In addition to these blood relatives, the household included several slaves or servants and six to eight hired hands (boarders), plus a steady stream of visitors who extended the numbers and experiences of the family. It was not uncommon for fifteen to twenty people to be fed in the dining room on a given day.[4]

The Homestead

From 1839 until he was impaired by illness in the 1880s, Waltus Watkins would shape his physical environment. Modern visitors to the Watkins Mill historic site gain awareness of what analyst J. B. Jackson has termed the "vernacular landscape," one of "two very different but complementary elements in our landscape; one established and maintained and governed by law and political institutions, dedicated to permanence and planned evolution; the other the vernacular landscape, identified with local custom, pragmatic adaptation to circumstances, and unpredictable mobility.... The beauty that we see in the vernacular landscape is the image of our common humanity: hard work, stubborn hope, and mutual forbearance striving to be love."[5]

The farm was designated "Bethany" in 1842, a name with both religious and secular implications. Watkins may have envisioned a place of refuge like the biblical Bethany, where Jesus periodically retreated for refreshment at the homes of his friends, or a place of sophistication like Bethany College, founded 1840, in his native Kentucky, or perhaps both. The name was also in harmony with the name of the community cemetery, Bethel, located two miles to the south.[6]

The Cabin

Watkins, perhaps with help from neighbors, erected a small (18 feet by 23 feet) two-room cabin for his family in section 10 of Washington Township. Investigators disagree about the construction date for this cabin; some believe that it was built in 1842 and that Watkins may have had an earlier cabin somewhere in section 15. The structure was a variation of cabins built throughout the Mid-Atlantic and Upper South regions, with rough-cut logs, flat on four sides, forming corners in half-dovetail or V notches. The larger room (16 feet by 16 feet) had a fireplace and served multiple functions, while the smaller, narrower room was equipped with a wood-burning stove and served as a kitchen. Sleeping quarters were provided in an enclosed lean-to across the back of the cabin, as well as in a small loft. A porch stretching the length of the front mediated between the indoor and outdoor space and their separate activities. In sum, the Watkins home was traditional in its external appearance and its internal functions.[7]

The cabin proved more than temporary quarters. The family lived there perhaps fifteen years; by the time Watkins moved them to a more commodious structure, there were between ten and thirteen people living in the cabin.

The Big House

As early as 1845, Watkins contemplated building a house to replace the log cabin. When brother-in-law William B. Holloway of Harrodsburg, Kentucky, was approached about the project, he replied, "I would like very much to build it for you I would judge as Polk is elected you will want a right fine house." But Watkins had made no specifications, and Holloway felt obligated to fulfill his commitment to continue laboring for John Handy. The house would be made from brick, and thus Watkins ventured into building kilns, digging and weathering clay, experimenting with recipes, and harvesting wood to stoke the fires. Waltus's son John reminisced, "Pa must have made enough bricks to build the town of Lawson before he got the hang of it." Construction commenced in 1850 but was not completed until 1854. Waltus's sister Mary wrote in 1852, "How do you come on with your new house? I hope I may yet get to live to see you all situated comfortably in it. I know you act wisely in building & remaining where you

Illustration of Bethany home. From the *Illustrated Historical Atlas of Clay County, Missouri.*

Illustrations of other county farms emphasized—often from a bird's-eye vantage point—the numerous support buildings that generally surrounded the main house. In the case of the Watkins residence, the artist chose an angle that rendered invisible any service building other than the smokehouse, which peeks out from behind the house and is largely obscured by a tree. Watkins, like fellow southerners, probably built such structures in his backyard. Details in the sketch, such as the pump and orderly tree plantings, have been confirmed archaeologically. The closed shutters on the northeast corner of the house and the location of the fruit orchards close to the house on both the west and south have also been corroborated. The drawing gives no clue as to nearby gardens or brick kiln. The details provided may be accurate, but the artist's chosen perspective limits the usefulness of the drawing to historians interested in the actual layout of the farmstead.[a]

[a] Rucker, "Archaeological Investigations."

are, instead of going away to Oregon." Although by then John Handy was no longer in the construction business, he undoubtedly commented on Watkins's project during family visits or one of their periodic joint ventures in driving mules from Missouri to Kentucky, and he continued to solicit craftsmen to go to Bethany.[8]

How Watkins positioned his new home (known within the family as the Big House) and what design he followed were symbolic actions. Geographer D. W. Meinig has noted that as the American empire expanded at this time, men like Watkins manipulated landscape "to create symbols of authority so as to induce a shift in the psychological focus ... and gain allegiance not simply from fear but from a respect and even cultural attainments and superiority of the ruling power." Watkins chose the highest ground in his landholdings—a thousand feet above sea level—and faced his house to the east, thus providing a commanding view of the near approach of the Lane (the road leading from the town of Greenville to the house and mill), as well as of the Richmond-Plattsburg road. (As a practical matter, the proximity of the well he had dug in 1846 to serve the log cabin may also have affected his decision.) At this time, his purview encompassed his gristmill, sawmill, livestock holdings, orchards, and sections of his pasturage. He erected a series of buildings in the hollow just northeast of the house and also to the west, in back of the house, to support the family's commercial ventures and domestic needs. Subsequently he would further shape the land—sculpting roadways, erecting fences, digging ponds, cutting woodlands, and planting—to accommodate his various undertakings and further reflect his social standing.[9]

Historian Richard Bushman asserts that pursuit of gentility by families like the Watkinses "was worldly not godly, it was hierarchial not egalitarian, and it favored leisure and consumption over work and thrift. These values ran at cross-purposes with religion, republicanism, and the work ethic, powerful complexes of values subscribed to by the same people who wanted to be genteel.... Refinement held out the hope of elevation from ordinary existence into an exalted society of superior beings."[10] The Greek Revival brick home erected on the Bethany hillock certainly symbolized refinement to its inhabitants, as well as to all who ventured there. Yet the new two-and-a-half-story home also evinced certain aspects of vernacular architecture. Construction materials such as limestone foundation blocks, bricks from Watkins's own kilns, and woodwork fashioned from trees on site, indicated local self-sufficiency. The "common bond" pattern for the brickwork did not require much bricklaying skill. The modified center hall plan cannot be clearly traced to any of the mounting number of pattern books then being published. Rather, it can be inferred that Watkins fashioned his own plans, probably relying on the counsel of John Handy and the experience of his builders. For example, the carpenters squared the normally cylindrical pillars in the front of the house, upon which Watkins placed an Italianate iron railing for the top of the porch. Moreover, they used building techniques such as mortise-and-tenon joints in the heavy wooden beams, an intermediate step between the crude and labor-intensive notches cut for the log cabin and the balloon-frame structures popular later in the nineteenth century.

The home's furnishings further signify the family's quest for refinement. As guests passed through the front door, they were welcomed and ushered into the formal parlor on the right. Bushman states, "The crucial characteristic of a refined house was a parlor, free of work paraphernalia and beds, dedicated to formal entertainment and the presentation of the family's most decorative possessions." The parlor, lacking any economic function, signified the family's role as consumers, with its genteel and polite ambience symbolized by the carpet on the floor and the family piano. Watkins bought all new furniture for his parlor in 1854 and added a piano in 1870. Although heated by a stove, the Watkins parlor hearkened back to the traditional family hearth by retaining a mantel.[11]

As symbolic as the parlor was the semicircular, semi-floating staircase that led from the entryway to the second floor. Legend has it that working the native walnut required the skill of craftsmen trained in the East and delayed completion of the house for two years. The existence of five additional staircases allowed for segregation of guests from family, parents from children, brothers and male boarders from the girls, public functions from private. Such stairs also provided access, such as to the basement food storage areas for those preparing meals.[12]

Outbuildings

Milking and cow barns, plus a silage pit, were located southwest of the house. The big timber-frame mule barn, measuring perhaps 30 feet by 90 feet, was to the north. Therein were the stalls for the driving

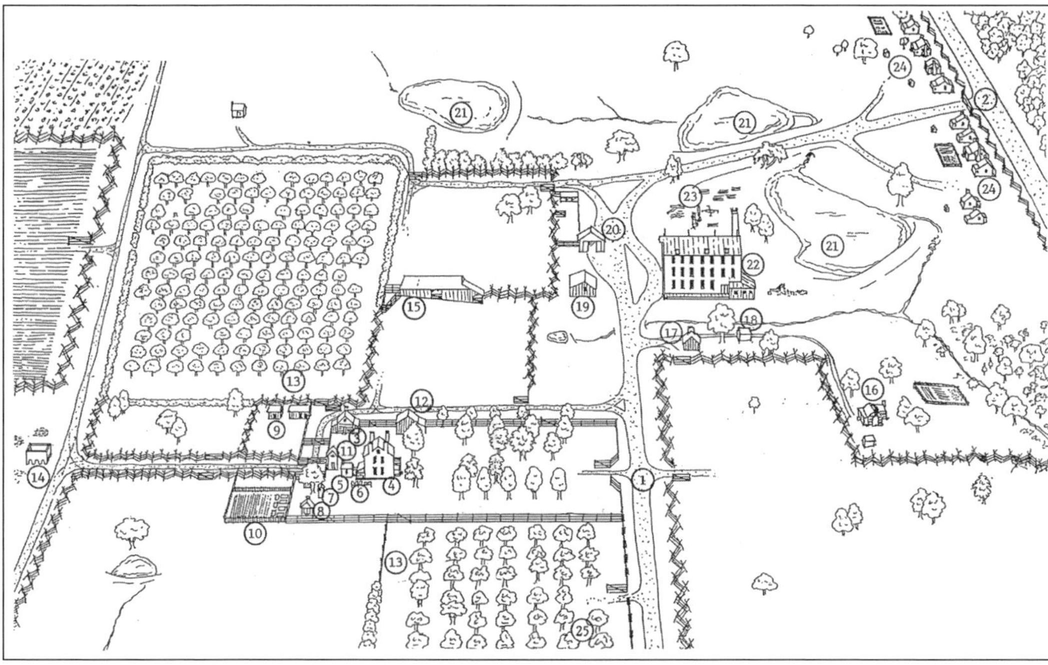

Homesite layout. DNR, WMSHS, drawn by A. Matthews.

1. Lane
2. Bethany Road
3. Log Cabin
4. Brick House
5. Summer Kitchen
6. Beehives
7. Privy
8. Fruit Dryhouse
9. Chicken House
10. Garden
11. Smokehouse
12. Icehouse
13. Orchards
14. Brick Kiln
15. Mule Barn
16. Waers's House
17. Blacksmith Shop
18. Bleach House
19. Woolshed
20. Scale House
21. Mill Ponds
22. Woolen Mill
23. Sawmill
24. Smokey Row
25. Watkins Cemetery

mules, along with two large corn cribs that were served by wagons entering and exiting through a central throughway. A lean-to on the eastern end of the structure may have housed the farm's workshop. A large catchpen surrounded by a 7-foot fence expedited special handling of the mules. A 5-foot-high board fence kept large draft mules within the pasture and away from the house.

The sawmill with its circular sawblade (another example of Watkins's penchant for advanced technology) was located across the access road from the barn. With harnesses for as many as eight horses or mules, it was not dependent upon available waterpower and could therefore be used in any season.

To build these structures and repair machinery, Watkins relied not only on his own training, but also on the talents of a hired blacksmith. The blacksmith shop, though small (16 feet by 16 feet), was significant in Watkins's transition to the industrial age. The smith was the crucial mechanic at the farm; he made and repaired tools, farm implements, household utensils, and hardware. In addition, either he or a farrier shod the considerable number of mules and horses. Many blacksmiths were also wheelwrights, whose repairs kept Watkins's wagons and carriages in operation. Watkins's later smiths repaired mill machinery and occasionally created their own devices, such as the yarn bundling press. Thus the Bethany blacksmith shop was the equivalent of a modern auto repair shop, a welding shop, a farm implement repair service, a machine shop, and a hardware store, all rolled into one. Although the smith's services were obviously critical to Watkins's fortunes (and to the region's economy), census records show no smith in residence at Bethany.[13]

The cluster of buildings around the Watkins house signified Waltus's ascent in the region. The buildings were also the places where the inhabitants exchanged produce and services and shared experiences. When newly arrived cousin Dr. William H. Cosby, for example, married Miss Eliza F. Porter in March of 1846 at the New Hope brick church, it was reported that "between 7 and 800 were disposed to be present." The wedding party, including ten attendants and all close relatives, was hosted by Waltus and Mary Ann with a bountiful feast at Bethany. Similar festivities the following day were held by other cousins, the Gills, who lived two and a half miles north on the Richmond–Plattsburg road.[14]

DAILY LIFE

Relentless labor filled the days of the pioneering Watkins family. The work was usually difficult, often tedious, and sometimes dangerous. Trees were felled, wood hauled, stones dug, sod broken, crops produced and harvested, butter churned, food preserved, and meals prepared—all largely by hand. Though tools and animals eased the burden, running the farm and household required a huge investment of human muscle power and energy.

At Bethany, tasks were largely divided by gender, yet the spheres of Mary Ann and Waltus never were totally separate. This was the common pattern, as historian Jack Larkin has observed: "Men's and women's tasks on American farms were intertwined and almost totally interdependent. But in space, time, tools and authority they were distinct. Farmyard, garden, house, kitchen and hearth, in diminishing concentric circles, enclosed and bounded women's daily realm. The world of men's work circled outward. It began with the farmyard, barn and workshop, and went on through gates and across fences to fields, pastures, woodlands and then to roads leading off the farm."[15]

The kitchen was a key nexus for Mary Ann's work, and was strategically placed in the Big House. Located in the northwest corner room of the first floor, just off the dining room, the kitchen had immediate access to the dairy and food storage cellars in the basement and to the water pump north of the back porch. In the warm part of the year, food preparation was moved to the separate brick summer kitchen adjacent to the back porch. At the cookstove and fireplace, Mary Ann guided her daughters, slaves, servants, and perhaps female kin and visitors, as they labored tediously to prepare the daily meals and to process and preserve food for future use and for sale.

In addition to food handling, child care, and general housekeeping, the women's chores included sewing family clothes and weaving rugs.[16] That Bethany ever came close to being self-sufficient was traceable to the talents and energies of the domestic workers laboring under the watchful eye of Mary Ann Watkins.

The focus of Waltus's domestic responsibilities lay outside the house, in the farmyard and beyond. There he supervised his sons, slaves, and hired hands in toiling with the animals, buildings, machinery, farmland, orchards, and woodlots that supported the household operations. His work and Mary Ann's frequently intersected, especially where food production for the family expanded into commercial ventures.

Food for Family, Food for Sale

The Watkinses spent a large part of the spring, summer, and fall making sure they had enough food put up to see their large household through the bare winter months. Any surplus was sold or bartered for other goods. This blending of domestic and commercial activities illustrates one response of Americans such as the Watkinses to the market revolution. Historian Daniel Walker Howe has offered this appraisal: "The ways of the market were not perceived in sharp opposition to customary rural ways. There were means of combining and synthesizing them.... The market brought people new power to shape their own identities."[17]

Milk from the Watkins herds was chilled by ice that had been harvested from local ponds and stored in sawdust in the icehouse until needed. Some milk was made into cheeses, which aged on basement shelves. The milk also yielded its cream to be churned by hand into butter. Cheese and butter not only satisfied the household's appetites but were also part of the regional market for barter or for cash sale. Mary Ann supervised the dairy operations, keeping vigil over the temperature and humidity within the cool whitewashed limestone walls of the basement dairy rooms and thereby monitoring and controlling the quality of the various products.

Produce from extensive orchards and garden plots provided ingredients for Mary Ann's recipes, herbs for Waltus's botanical medicines, and income from sales to neighbors. In the garden, perhaps an acre in size, vegetables were grown in raised beds; potatoes and other row crops in large patches; and fruits, herbs, and flowers (roses, lilacs, and white lilies) along the borders. Potatoes, cabbages, and other root

Among family archives is a leather-bound cookbook kept by the youngest Watkins daughter. Carrie was 16 and already an accomplished cook when she started this personal and handwritten record, later supplemented by newspaper clippings, in a journal previously used to record school district meetings. It was largely comprised of dessert, pickle, and condiment recipes, many of which only listed ingredients, with minimal or no directions. From such omissions, it can be inferred that Carrie, from repetition, knew the measurements and procedures and how to gauge the inconsistent temperatures of the Dutch ovens on the hearth or the wood-burning stove. At some point, she pressed two four-leafed clovers in her cookbook, perhaps to ensure the success of her efforts. She also had the habit of recording the levels of her likes and dislikes. Lee Dye, an avid student of the cookbook who, like Carrie, has labored repeatedly over the fire and stove to feed laborers at the site, has concluded, "We find hidden dividends over and above the value of antiquity in Miss Carrie's recipe book. In it we find thrift, sensuality, and practicality."[a]

[a] An innovation introduced by Fannie Farmer's *Boston Cooking-School Cook Book* (1896) was the use of standard measurements; the book also listed ingredients first before providing preparation and cooking directions. Madeline Matson, a close student of Missouri foodways, has noted the popularity of cheaply printed cookbooks from 1800 onward. Matson, *Food in Missouri,* 123–24; Caroline Emma Watkins [recipe book, 1870–1910?], WMSHS; Dye, "When Eggs are Dear," 24.

crops were stored in barrels in the dark, cool cellar as staple foods for the winter months.

Peaches, cherries, apricots, pears, and apples were grown in orchards at the farmstead as well as on tenant farms. Watkins did not dabble in fruit. Characteristically, he entered the trade with both a scientific attitude and commercial aspirations. He experimented with thirty-five species of pears; in 1869 he purchased 610 pear trees.[18]

While Waltus selected fruit trees to be planted and supervised their cultivation, it was Mary Ann who oversaw the extensive drying and marketing of the orchards' yields. In the commercial-sized fruit dryhouse, fumes from flowers of sulphur mixed with smoke from a shallow pit fire to keep the fruit from turning brown and to discourage insects. Here, as in other food processes, skill and attention were crucial. The fire had to be kept just right—from 150 to 180 degrees Fahrenheit—and the whole process was carefully monitored by the Watkins girls over a period of a day and a half. The dried fruit was then put up in glass jars in the basement storage rooms. Even damaged or windfall fruit could be—and usually was—cut up and preserved or dried, thereby extending the orchards' productivity. The Watkinses used only one-third of the yield, selling the surplus by the case.

Livestock production also required the attention of both Waltus and Mary Ann and their respective helpers. Like many other rural western households of the nineteenth century, the Watkinses consumed two great staples: corn and pork. By the 1860 census, Missouri ranked second among the thirty-three states in corn production and

Figure 2.1: Census data on Watkins family products and holdings

	1850	1860	1870
Size of Watkins household	8 family 1 boarder	11 family	8 family 6 boarders
Value of household manufacturing	$50	$75	—
Value of animals slaughtered	$645	$1,500	$2,000
Value of farm products	—	—	$4,000
Honey	250 lbs.	300 lbs.	—
Molasses	—	300 lbs.	—
Orchard yield	$50	$400	$5,000
Cheese	100 lbs.	50 lbs.	—
Butter	300 lbs.	500 lbs.	500 lbs.
Hay	20 tons	30 tons	20 tons
Oats	600 bu.	150 bu.	2,000 bu.
Corn	4,000 bu.	3,000 bu.	8,000 bu.
Wool	120 lbs.	150 lbs.	150 lbs.
Value of livestock	$3,945	$2,000	$5,000
Implements	$300	$400	$1,000
Slaves	3	3	0
Improved acres	650	700	3,450
Unimproved acres	590	1,400	—
Total worth of property owned by Waltus Watkins	$15,000 ($1,160 in land)	$21,000 real $15,000 personal	$122,000 real $20,000 personal
White population of township	1,176	1,569	2,522
Black population of township	252	445	87

Source: Agricultural Schedules, Seventh, Eighth, and Ninth U.S. Censuses, Washington Township, Clay County, MO, 1850–70.

fourth in swine production. A typical farm family consumed a dozen "elm-peelers," "stump-suckers," or "land-sharks," as swine were variously called, every year. Hogs were more efficient than cattle or sheep in converting grain to meat, "for what is a hog, but fifteen or twenty bushels of corn on four legs." The Watkins family butchered from eighty to a hundred hogs each year right outside their smokehouse, where large kettles were set up for scalding the carcasses (to expedite hair removal) and for rendering lard. Waltus particularly favored Berkshire hogs in excess of three hundred pounds, as they yielded much lard for use in domestic cooking as well as in industrial lubricants. Hams and bacon, after salting for three weeks, were hung with sausages and smoked over slow hickory fires for another three weeks. They were then stored amidst ashes in the hollow sycamore log or "powder box" in the smokehouse.[19]

Beef and poultry were also in abundant supply, an indication of the family's affluence. The Watkinses, especially John, took great pride in raising Shorthorn or Durham cattle; as late as the 1880s, between five and six hundred head of Shorthorns ranged over their property. The Shorthorns "were valued for their size, early maturity, smoothness of flesh, high percentage of choice cuts, and, perhaps most importantly for Missouri, their tremendous prepotency."[20] Mary Ann Watkins kept anywhere from three hundred to six hundred chickens on the farm, and upwards of a hundred and fifty turkeys. The former provided not only eggs but also frequent fried chicken for the Watkins dinner table; the latter were fare for special occasions. In addition, the income from egg sales often provided critical cash flow for the family. The Watkinses also gathered honey and beeswax from hive boxes on the premises to provide sweets for their table. Surpluses were shipped in barrels for sale in Kentucky.

Watkins's hog raising led to a legal dispute with a neighbor in February 1847. His attorneys, Democrats of local renown, filed suit "that John Brammer had slandered Watkins by accusing him of larceny." Watkins, Brammer let his neighbors know, had stolen seven of his pigs and placed the Watkins brand on them, as well as on a sow Watkins allegedly gained from Mace Rust. Brammer ranted that Watkins should be "'put away' in the state penitentiary." Watkins's attorneys contended that "these statements were totally false, and had damaged Watkins's reputation in the amount of $10,000." The case was not called for more than a year. Finally, on 2 March 1848, the jury found Brammer guilty and awarded Watkins $7,000. The same day Watkins filed a remitter in the amount of $6,800, thereby leaving Brammer responsible for a $200 balance.[a]

[a]Waltus L. Watkins, plaintiff, versus John Brammer, defendant, Fifth Missouri Judicial Circuit, box 13, February and August 1847, March 1848, Clay County Circuit Court Records, Clay County Archives.

Sickness and Death

The Watkinses grew plants for medicinal purposes as well as for food. It was an era when epidemics were a fact of life, with cholera and ague (malaria) particularly rampant.[21] Standard medical care—consisting largely of bloodletting, blistering, and dosing—was as often fatal as curative, and the sectarian healers who competed vigorously with orthodox physicians were no better. Watkins condemned the widespread use of calomel (a strong purgative) "as the common enemy of mankind," yet he would probably agree with historian Larkin that "most [medical] practitioners were not charlatans and most patients were not fools. Within their horizons of understanding, they could believe, usually on rational grounds and often on the best evidence available, in the efficacy of treatments which now appear useless or even dangerous." Personally, Watkins was attracted to botanical medicine. He grew a variety of herbs, studied their systemic uses, practiced avocationally, and sought to recruit professional healers to his neighborhood.[22]

When medical care failed or old age took its toll, the funerals were most frequently held at home, as was then customary in rural America. In death, Waltus and Mary Ann symbolically kept their distance from their contemporaries, many of whom were laid to rest at Bethel Cemetery. The Watkins family cemetery was located south of the house, reportedly because matriarch Jane Minter Watkins wanted to be buried in the peach orchard (a wish that was honored when she died in 1862).

Today there are eight marked graves for members of the Watkins household; others may be unmarked. There is no record of the burial sites of the two children, John Holloway Watkins and Alfred Watkins, who died in 1839 and 1840, respectively. In the 1930s, Joe Watkins erected a stone wall to keep the cattle from wandering into the cemetery. He also replaced his parents' and sister's headstones, possibly damaged by cattle, to match that chosen for the grave of brother John Handy Watkins, who died in 1931. The old headstones were buried in cement under their replacements. A nephew of Mary Ann Watkins, William H. Holloway (died 1938), was the last to be interred at the plot.[23]

A Moral and Intellectual Inheritance

Education

A crucial chord in the Watkins family legend passed from generation to generation was the need to be well educated so that one may fulfill one's calling. Training at the farmstead under the parents' tutelage was emphatically reinforced with formal schooling outside the home. Perhaps as compensation for his own abbreviated formal education or perhaps as a measure of his accumulating wealth, Watkins financed higher education for his sons and daughters. Some of the older boys returned to Georgetown, Kentucky, for college; after the 1850s, younger brothers matriculated at William Jewell College. George traveled as far as Hamilton, New York, to study languages and the arts, while Adoniram Judson (also known as Judson or A.J.) ventured to Alabama and perhaps St. Louis for business school. As Watkins believed girls were "more dependent than the male part of Creation," he was especially attentive to providing schooling for his daughters. The Watkins daughters in succession gained the amenities of study offered at Liberty Female Institute, whose catalog proclaimed, "Whenever it becomes manifest that a pupil needs other influences than moral suasion and the honors of the Institute to incite her to industry and duty, she will be immediately, and, if possible, without disgrace, returned to her parents." Collections of correspondence among the Watkins girls, as well as some of their texts, indicate the drudgery, anxiety, and excitement of their school years. Lizzie, for example, was most proud of her 1868 graduation from Clay Seminary, then deemed the most prestigious female institution in western Missouri.[24]

The theme of education as supporting one's life role can be glimpsed in letters written to Waltus during his apprenticeships; in correspondence of Waltus's firstborn son, George, when he went away to college; and in Waltus's communication to his daughters attending school in 1864. In 1825, Benjamin Watkins lectured his eldest son that "self knowledge is more valuable than all the wealth of India." People should endeavor "to learn human nature." Each day, Waltus ought "take a few moments…to think what you are and what created for." Benjamin concluded this parental missive, "May the good Lord direct you through life's uncraven way is the desire of your unworthy father."[25]

In 1855, George Watkins was at Georgetown College in Kentucky. Though he was within support of the Handys and other kin there, he was homesick for Bethany. In letters to his parents, he observed, "It is at home affection unfolds her store of fireside comforts and it is there brothers & sisters constitute one unbroken chain of social fellowship in the fireside of serenity and peace." As he completed his first year at Georgetown, he reflected, "I love my parents and feel under many obligations to them…and there is no danger of my forgetting work for I served under you. When I see weak puny and efeminate [*sic*] boys here at college whose parent never required them to work I always feel thankful that I was strengthened by exercise." Perhaps telling his parents what they wanted to hear, George related that after reading Benjamin Franklin's life, he had discovered an "example i think worthy of imitation." George had earlier instructed brother John Watkins to study well at the Academy, to read and to memorize so that "when you get to College you will never regret that your youth was well spent in acquiring useful knowledge."[26]

In the midst of the Civil War, Watkins advised some of his daughters as they approached their final school examinations, "Do not think you will then be done with life's troubles for they will have just begun & I hope that you will then really perceive and know that you have only laid the foundation of an education." His concluding line to them echoed the advice from his own father: "May we all do our duty on Earth and be saved in his Kingdom is the wish of your Father."[27]

The Watkinses' moral and intellectual inheritance—gained at home, in school, and from their church—was passed down from one generation to the next. Waltus was especially sensitive to perpetuation of the family legend (supported by genealogical tracings formally begun in 1861) to bond his offspring together and to point out future obligations. The message was the importance of maintaining an upright and orderly role of leadership in the community. Not every child fit the mold. John Handy became somewhat of an iconoclast, while Catherine Jane (Kate) was an early suffragette and feminist. Jewell appears most wayward. Not only did he periodically confess transgressions at church, he ultimately wandered from his family and Bethany to Wyoming and Colorado. There, out of contact or control of parents, siblings, or wife, he alternated between his vocation as a dentist and avocations as a big game hunter and travel reporter.[28]

A "Clear and Bright Conversion"

In 1849, Waltus Watkins underwent a life-changing religious conversion. It was an event for which his relatives, particularly his sisters, had been praying for a long time, and it would shape his family, community, and commercial interactions from that point on.

An early correspondent from Kentucky in 1835 had observed of Watkins, "You either were or feigned to be skeptical in the important and vital concern of the Christian Religion."[29] Such hesitancy vexed the deeply religious Mary Watkins Handy. While John Handy sustained Waltus's enterprises, Mary prayed in each letter for her brother's soul. Sister Tabitha also beseeched Waltus to pray nightly for his conversion and pointedly noted that the truly converted must reject worldly possessions. In 1847, Mary used the death of their sister Caroline to convince Waltus to "try and become a Christian & love & serve the Lord."[30] However, it was not until August 1849, at age 42, that Watkins joined the Baptist Association at New Hope Church. According to yet another sister, Rebecca (who lodged with him for extensive periods in the 1840s and 1850s), Watkins had finally yet wholeheartedly responded to one of the periodic crusades that swept the region.[31]

In the 1820s, when Clay County was the frontier of both Missouri and the nation, religious institutions were looked upon as the initial benchmark of civilization. Before the emergence of towns or the erection of schools, churches were organized. The Methodists had established themselves in Watkins's neighborhood in 1822 and the Baptists in 1829. New Hope Baptist Church in northern Clay County was served by a series of gifted but plain preachers. As historian Robert Handy has pointed out, "Throughout the week such a man would normally earn his living as did other members of the flock, but on Sunday, making up by sincerity what he lacked in formal training, he exhorted his congregation, using images and idioms fully understandable to the people." For the first six years, Ben Riley and Howard Everett alternated in the pulpit. Then Soloman Kimsey was named pastor. Prominent revivalist Moses E. Lard later remembered, "On Sundays, the countryside flocked to church—the wags and wits to swap horses and whistles, and to bet on coming races, and the Christians to hear the sermon and to relate their experiences as was fitting.... Solomon Kimsey always preached the same sermon and it always had the same effect—that is, the women were left crying and the men discussing the election." Within two years, Kimsey had sundered the congregation over "matters of communion" and "too much commercialism" in his preachings. It was the highly reputable Robert James's task between 1844 and 1850 to revive the membership and build a new church for New Hope. James had superior training, including attendance at Georgetown College in his native Kentucky, and energy that led to organization of the North Liberty Baptist Association in 1844. It was he who in 1849 would finally win over Waltus Watkins.[32]

The revival began in the middle of July and lasted throughout August. Presbyterians, Methodists, and "old baptists" alike were affected. Three ministers, including James, exhorted the sinners to convert, with the result that forty-two made the decision to be baptized. Robert James reportedly "said he never saw as many at one time before. It looks like the good Lord has begun a good work there." Waltus professed his faith and was accordingly accepted into the church on 7 August. Sister Elizabeth Carter related how Waltus, after being quite sick for the previous three weeks, responded to preacher James's call: "[The congregation saw] Waltus' deep conviction...his clear and bright conversion while on the mourners' bench, and then, in the course of a half hour, when the invitation was given, to see him come out so humbly—though boldly—and relate his experience." Three of his sisters rushed to embrace him as he cried out for mercy. Preacher James told one of Waltus's sisters, "I just wish your mother was here this evening. I know the old lady would have shouted." It was impressive, especially for those who had prayed for nearly two decades for such an event. Rebecca announced, "He says he would not take the world for the hope that he has." Their brother James, however, resisted conversion.[33]

Waltus "is a changed man," reported his sister. "He is regenerated and born again. His whole theme is to talk about religion, go to

An outward indicator of Watkins's commitment may be seen in the names he gave the children born in the decade after his conversion. Two of his last three sons bore the names of religious leaders: Waltus Jewell (born 1852), for the benefactor of William Jewell College, and Adoniram Judson (1856), after the noted Baptist missionary to India. Daughter Caroline (1854) was named for the sister whose death had been used by the family to prompt Waltus's religious conversion.

meetings and read the Bible." Though there are hardly any records of Watkins's religious ponderings, the balance of his life offers powerful testament of devotion to his faith and his church. His baptism was a major benchmark in his life. Not only did he feel his soul regenerated, he also redoubled energies in civic and capitalistic pursuits. Henceforth, "his duties to society, morality, religion and his financial obligations were his supreme law." Watkins's faith provided a means to explain both his personal successes and the failures of some of his peers. The core of Watkins's evangelical beliefs provided an ideological vantage point, a way to perceive the world and explain his own place in it and his destiny. In eulogizing Watkins at his death, preacher W. C. Barrett, proclaimed, "As a Christian he weighed well his thoughts, and fairness and stability characterized his actions; he was successful in business and liberal to a fault. The various religious institutions of the day shared in his benevolence; the mission cause home and foreign."[34]

Folk such as the Watkinses gained a large measure of how they oriented themselves to their world from their religion. The ethic of the Industrial Revolution—individual enterprise, moral discipline, and success—was reinforced by such evangelicalism. Moral and economic virtues were complementary. The inner grace of regeneration was to be declared in dramatic and emotional proclamations. For Watkins and other community leaders, such experiences delivered them from a sense of guilt that might arise from personal material quests. Personal liberation from sin was the complement of social control.[35]

A "benevolent empire" was the goal of numerous activities supported by Watkins and fellow reformers of the era. Through Sunday schools, missionary endeavors, and Bible studies, they sought to liberate victims from their own passions. Historian William Barney has observed, "By transforming liberation into control and secular progress into [the] coming of the millennium, evangelicalism exerted a powerful appeal to the middle class. For all its boundless optimism, evangelicalism promised a society of self-regulating individuals who, by submitting to God's moral authority, set limits on their own lust for power." In the Industrial Revolution, managers and workers alike could thus gain self-improvement amidst social harmony. Social control took precedence even over moral uplift.[36]

Avoiding "The Intoxicating Bowl"

Waltus Watkins, as a community builder, was sensitive to the greatest reform crusade of the 1830s, the temperance movement. Between 1790 and 1830, alcohol consumption in America had increased 200 percent; by Watkins's era, per capita consumption was in excess of seven gallons of pure alcohol per year (roughly three times the present-day rate). Drinking spirits was accepted as a normal part of life. One analyst has noted, "Drinking was an accepted part of Upland Southern culture; Southern religion did not effectively proscribe it, and efforts to root it out were too closely related to unpopular Yankee reform movements. Temperance had some appeal to an industrializing middle class, including some of the Southern born. It had little popularity, however, in the less developed parts of these Midwestern states, dominated by the Southerners." As the new market economy took hold, temperance (with its more stringent strategy, prohibition) was aimed at correcting the societal disorder of which alcohol consumption was the most prominent symptom. For Watkins and others like him, the drinker was viewed both as a sinner and as a threat to society.[37]

Watkins became a most prominent advocate of temperance. He reportedly enrolled in the Washington Temperance Society, a nationwide group that between 1837 and 1843 developed a self-help program among Protestant workers. From the time he took the pledge until his death, Watkins was enthusiastically proud to act as a model for a well-ordered society. In 1870 he reflected, "I have not for more than thirty years used calomel or ardent spirits—nor have I for more than twenty years used any coffee or tobacco—though I once used all of them." Watkins exhorted his stepnephew, John Luke, in 1844 to join a temperance society at "the first opportunity the best way of keeping clear of drunkenness to never indulge in the intoxicating bowl." Abstention from alcohol was chief among the qualifications needed to work at Watkins Mill, as the Watkinses, like most factory owners, believed drink and machines were a dangerous mix.[38]

STAYING PUT

Waltus Watkins lived in restless times, a fact of which he was frequently reminded by the westward movement, the waves of visiting kin, and the periodic dispatch of his children to further their educations or careers elsewhere. Social historian John Mack Faragher has depicted the era as a "story both of a transient majority, the people called 'movers' by contemporaries, and those men and women who persisted in the area and put down roots.... The private histories of the mobile majority are largely lost to us. The recoverable histories of those who persisted, on the other hand, tell us much about the importance of traditional family and kinship patterns on the moving frontier." Watkins was one of those who persisted.[39]

By 1846, Waltus was recognized as the most prosperous family member in Missouri, the one whom other family members (particularly his sisters) petitioned for assistance in raising their families. Still, he seems to have been susceptible to Oregon fever; he subsidized a reconnaissance of Oregon by his colleague James M. Horn and wrote to sister Caroline, "I have a large farm here and no hands of my own to work it so I rent out most of it. I now have 640 acres under fence and 450 under cultivation—120 in corn and 300 rented out for a third of the crop of corn, wheat, and oats. No hemp is sowed on any of it this year. Since my farm is so large and so much trouble, I have lately concluded to sell it if I can find a purchaser. I will sell low so can't you send me one? My land is as rich as that in Woodford and Fayette [Kentucky] with good waters and plenty of it in every field and pasture on the place." Some of his sisters were not enthusiastic about the idea of Waltus moving to Oregon. Mary's position was clear as she reflected on Waltus's "sly idea of going to Oregon...I hope it is a joke, for I think you are far enough from us now." When ultimately Watkins resisted the itch to move. Elizabeth rejoiced, "For all his land, good orchards, good water and plenty of everything but house room, he has too good a home there to leave, and it would be the rankest folly to think of going to Oregon." In short, Waltus Watkins the entrepreneur recognized when some risks were too big.[40]

Commentator Peter Drucker has observed that "the entrepreneur always searches for change, responds to it, and exploits it as opportunity." Watkins's multiple endeavors in western Missouri illustrated time and again such an aptitude. Having decided to stay put, he intensified his strategy of diversification and in so doing helped pull the area from colonization status to one of further development.[41]

Notes for Chapter 2

[1]Waltus L. Watkins to Caroline Hardin, 12 September 1846, WMSHS; De Voto, *Year of Decision,* 238ff.

[2]This fertility can be contrasted with statistical averages for children born to white women in Mary Ann's century: 7.04 in 1800; 5.42 in 1850; and 3.56 in 1900. Barney, *Passage of the Republic,* 75–76.

[3]Barney, *Passage of the Republic,* 71; Watkins Family genealogy charts, Research Files, WMSHS.

[4]"Household space in the log cabin was a scarce resource for which there were many competing uses. The large number of people who shared this space had to adjust their behavior to one another." Welch, "Silent Testimony," 4.

[5]Jackson, *Discovering the Vernacular Landscape,* xii.

[6]For references to Jesus at Bethany, see Matthew 21:17, 26:6; Mark 11:11, 14:3; Luke 19:29; and John 11:1, 12:1.

[7]Bray, *Archaeological Investigations;* Welch, "Silent Testimony," 2–6.

[8]William B. Holloway to Waltus L. Watkins, 23 March 1845, Mary Watkins Handy to Waltus L. Watkins, 5 July 1852, WMSHS; Roney, "Old Watkins Mill," 8; John C. Handy to Waltus L. Watkins, 27 February 1852, WMSHS; Mary Handy to Waltus L. Watkins, 2 July 1852, in Means, "Edited Watkins Family Letters," 124–26.

[9]Meinig, *Shaping of America,* 1771.

[10]R. L. Bushman, *Refinement of America,* xvi–xvii, xix. These values were passed on to the Watkins children; as Judson once reflected to John, "You know we were always taught to have nothing to do with other boys, with the impression that we were better than anybody else. That may do well enough at home but it will not work when one is away from his home thrown among strangers. When we get away from Mt. Vernon neighborhood we find some people who think themselves above us and they have something as well as us." Judson Watkins to Mary Ann Watkins, 4 March 1877, WMSHS.

[11]Welch, "Silent Testimony," 5–9, 12; R.L. Bushman, *Refinement of America,* 119–20, 251–65. Somewhat contrary to Bushman's model (283) was Watkins's placement of his bookcase. Rather than exhibit his readings to those the family entertained, his collection was placed in the family parlor. In his practical sense, books were to be read and reread rather than merely displayed.

[12]Spencer Holloway observed to Mary Ann Watkins (5 June 1854, WMSHS), "I was glad to hear that your house was nearly finished, when that is done you will be done with one more trouble.... I suppose Mr. Watkins is farming largely this year as farming prospects is somewhat flattering." Three years later Waltus's brother James constructed a similar dwelling, to be named "Prairie View." The interior was nearly identical to Bethany but the exterior, most notably with two-storied columns on the front portico, aspired to be more imposing. "On Building of the Watkins Home," Research Files, WMSHS; Means, "Other Watkins Also Settled Clay," *Clay County Sun Chronicle,* 8 September 1993.

[13]Bray and Yelton, *Archaeological Investigations,* 17–18, 55; Henderson, "Reconstruction of the Smithy," 3–9.

[14]Roney, "Historic Names."

[15]Larkin, *Reshaping of Everyday Life,* 17.

[16]For over thirty-five years, from the birth of her oldest in 1835 until her youngest went off to William Jewell College in the late 1870s, Mary Ann had the care of at least one minor child of her own, if not two foster children.

[17]Howe, "Market Revolution," 265–66.

[18]Waltus L. Watkins to [?], 1869, WMSHS.

[19]McCandless, *History of Missouri,* 48–51; Faragher, *Sugar Creek,* 64–66; William Cronon, quoted in M. R. Smith and G. Clancey, *Major Problems,* 209.

[20]McCandless, *History of Missouri,* 49. After the Civil War, Waltus and his son John improved their registered stock of Shorthorn cattle with a primebred bull they named "Horace Greeley" in 1869, after the New York newspaperman and politician.

[21]A nickname for some antebellum Missourians was "pukes," especially those living in fever-prone river bottoms. Whether this name was due to their rate of illness, their use of "puking" as a medical remedy, or some other cause is not known.

[22]Waltus L. Watkins to W. H. Crosby, 11 February 1845, WMSHS; Larkin, *Reshaping of Everyday Life,* 90. Fellow Democrat, temperance man, and botanic physician Thomas H. Freeman was a possible recruit. He wrote Watkins in 1844, "I hope you will still persevere in the good work & if you have not faith encourage embark in the best of all cases, do all you can in favour of morals & humanity." Freeman to Watkins, 4 October 1844, WMSHS.

[23]"Watkins Mill Guidebook," WMSHS, 25.

[24]Waltus L. Watkins to My Dear Children, 3 June 1864, WMSHS; Ethel Massie Withers, "Education in Clay County," Research Files, WMA Collection, WMSHS; commencement program, Liberty Female College, 1872, WMSHS.

[25]Benjamin Watkins to Waltus Watkins, 14 June 1825, WMSHS.

[26]George Watkins to Waltus and Mary Ann Watkins, 27 February 1856, 13 June 1856, George Watkins to John Watkins, 1 February 1856, WMSHS.

[27]Waltus L. Watkins to My Dear Children, 3 June 1864, WMSHS.

[28]The waywardness and brashness of Watkins's namesake (also a founding father of Craig, Colorado) is best discovered in Crockett, "Frontier Dentist."

[29]Thomas M. Freeman to Waltus L. Watkins, 13 January 1835, WMSHS.

[30]Mary W. Handy to Waltus L. Watkins, 14 January 1849, WMSHS. The sisters' shared religious enthusiasm is evidenced in a letter that Tabitha wrote to Mary and Elizabeth in 1844: "Though we three are attached to different branches of the militant church I hope that we have our names registered in the Lambs book of life and will all be found having on the wedding garment even Christ's righteousness." The evangelicalism of their faith, which often emphasized emotion over intellect, displayed a cluster of assumptions with a decidedly middle-class base: that the individual was competent to gain or to lose salvation; that society must strive to eradicate sin; that saved Christians, should their efforts succeed, would be rewarded with a thousand years of peace and progress. Tabitha W. Gill to Mary W. Handy et al., 8 March 1844, in Means, "Edited Watkins Family Letters," 83–85; Barney, *Passage of the Republic,* 97–104, 182–83.

[31]Rebecca Watkins to Dear Mother and Sister, 7 August 1849; Elizabeth Carter to Susan W. Hardin, 7–9 August 1849, WMSHS. Missourians at this time were stirred by evangelical denominations "in great numbers," observes historian Perry McCandless, "because of their less formalized creed and ritual, greater democratic organization, willingness to emotionalize the personal experience with God, and offer of quick salvation and immediate membership in the church." McCandless, *History of Missouri,* 206.

[32]Handy, *History of the Churches,* 167; Potts, "Waves of Revivalism"; Roney, *From Entry Fee,* 28–30. Robert James was highly reputable, but not so his sons, Frank and Jesse, whose outlaw careers are well known through both research and popular legend.

[33]*History of Clay and Platte Counties,* 153, 274, 281; Rebecca Watkins and Elizabeth Carter to My Dearest Mother, Sister, and Daughter, 7–9 August 1849, in Means, "Edited Watkins Family Letters," 113–18.

[34]Waltus Watkins obituary, from *History of Clay and Platte Counties,* 506; Mount Vernon minute book, c. January 1884, WMSHS.

[35]Kohl, "Social Control."

[36]Barney, *Passage of the Republic,* 101.

[37]Goldfield et al., *The American Journey,* 1:399; Rorabaugh, *Alcoholic Republic,* 232–33; Etcheson, *Emerging Midwest,* 60.

[38]*History of Clay and Platte Counties,* 506; *Liberty Tribune,* 11 November 1870; Waltus L. Watkins to John Luke, 2 July 1844, WMSHS.

[39]Faragher, *Sugar Creek,* 52, 60.

[40]Waltus L. Watkins to Caroline Watkins Hardin, 12 September 1846; John C. and Mary Handy to Waltus L. Watkins, 27 February 1852; Elizabeth Carter to [?], Spring 1851; Mary Handy to [?], [Spring 1851?], WMSHS.

[41]Peter Drucker, quoted in Pusateri, *History of American Business,* 8. In addition to opportunity, Watkins found that change brought risks and responsibilities. Close to home was the example of the Reverend Robert James, who had been enticed to California in 1849; when James fell victim to disease soon after landing in the goldfields (leaving young Jesse and Frank fatherless), Watkins stepped forward to assume some of his financial obligations. Watkins's own investment in the gold rush—in 1849 he subsidized James Horn's trip to California with $164 against the pending sale of a jointly owned sawmill—failed to produce appreciable dividends (and the settlement of the sawmill netted a $123 loss). Another example was Watkins's brother-in-law, John R. Hardin, "who was killed last year by the Indians [in Oregon] he has two children in the county one of them lives with me James M. Watkins is the guardian of their property." Waltus Watkins to James M. Horn, 2 June 1854, WMSHS.

3

Good Judgment and Attention

One of Waltus Watkins's primary strategies for improving his family's circumstances was land speculation. Throughout his lifetime, as government policies and his own economic resources permitted, Waltus Watkins seized opportunities to secure a large share of the available lands in western Missouri. He was doubtless in agreement with Missouri Congressman James B. Bowlin, who emphasized, "If you wish to preserve and perpetuate [the nation's] democratic form, you must pursue a policy tending to disseminate the lands amongst the largest possible number of people in the state." This was the fundamental creed of the Jeffersonian political economy, which Watkins endorsed and under which he prospered. Some of his land purchases were acquired for cultivation by his sons; some would be worked by tenants or agents; and some were held in hope that, with or without improvement, the price would increase and profits would be obtained. He may also have had in mind the role that land ownership could play in shaping the emerging community. A eulogist later claimed of Watkins, "This enterprising pioneer entered six thousand acres of land from the Government, and induced good men to come out and locate in the new country. He sold them land at very low figures, his object being to secure a good class of settlers in the community." While the eulogist exaggerated the magnitude of Watkins' landholdings and simplified his method of acquiring land, he accurately conveyed the purpose of Watkins's speculations.[1]

Land Acquisition

National Land Policies

Early American national land policy, established in 1785, was marked by two major characteristics. First, the square grid survey, advocated by Thomas Jefferson, ignored waterways and other natural contours of the land that traditionally served as boundaries. Watkins and fellow investors at the land office in Lexington, Missouri, gained deeds that specified the exact limits of the commodity they purchased, thus avoiding the

Missouri, 1877; Clay County is shown in gray. From the *Illustrated Historical Atlas of Clay County, Missouri.*

kinds of lawsuits over metes and bounds common in Kentucky and other areas surveyed in a more traditional mode. The grid, now most visible to air travelers over the Midwest, provided a ready, reliable, and inexpensive marketing device for the government's primary asset.

Second, the original policy's uniform price and minimum size of acquisition—originally set at $2 per acre for a section of 640 acres—tended to favor large speculators and overlooked wide differences in the quality of the land. In the first half of the nineteenth century, western states, with Senator Thomas Hart Benton of Missouri in the vanguard, succeeded in pressing Congress to liberalize land policies and adjust the variables of price, size, and credit to promote the rapid settlement of the public domain. By the 1830s, when Watkins entered the land market, the price of the national lands had been reduced to $1.25 per acre (compared to $5–$6 being reported for private transactions in

Liberty), and the minimum acquisition of federal acres had been reduced from 640 to 40 acres. With the Preemption Act of 1841, Congress made permanent a policy it had followed during the previous decade: squatters who settled on unsurveyed federal holdings were guaranteed the right to purchase up to 160 acres at $1.25 minimum price when the public auction was held. Then in 1854, a federal graduation act further adjusted downward the price of public land to as low as 12.5¢ per acre.[2]

Watkins's Land Purchases

In the course of his career as land speculator, Watkins acquired close to 5,000 acres through purchases and federal land grants. He started in Liberty with a 1.5-acre plot on "Kanzas Street opposite to Lewis Scott's bark mill," purchased in 1831 for $70 (and later the site of the nearly ruinous mill fire). In the spring of 1839, he chose an 80-acre plot in Section 15 of Washington Township as most suitable for settlement and promptly purchased several adjoining properties to expand his home base. Between the censuses of 1850 and 1860, he added 50 acres of improved lands and an additional 810 acres of unimproved lands to his investments, primarily prairie lands gained by federal grants in Ray County. Over the years, he increased his holdings in other parts of Clay, Ray, and Platte Counties. Depending upon the location, quality of soil, degree of improvement, and date of purchase, Watkins paid between $3.75 and $48.33 per acre for parcels ranging in size from 13 to 400 acres.[3] At its zenith, Watkins's property holdings totaled 3,660 acres (as recorded in the 1880 census), with the home farm comprising 1,200 acres. From his original 1839 investment of $300, he would

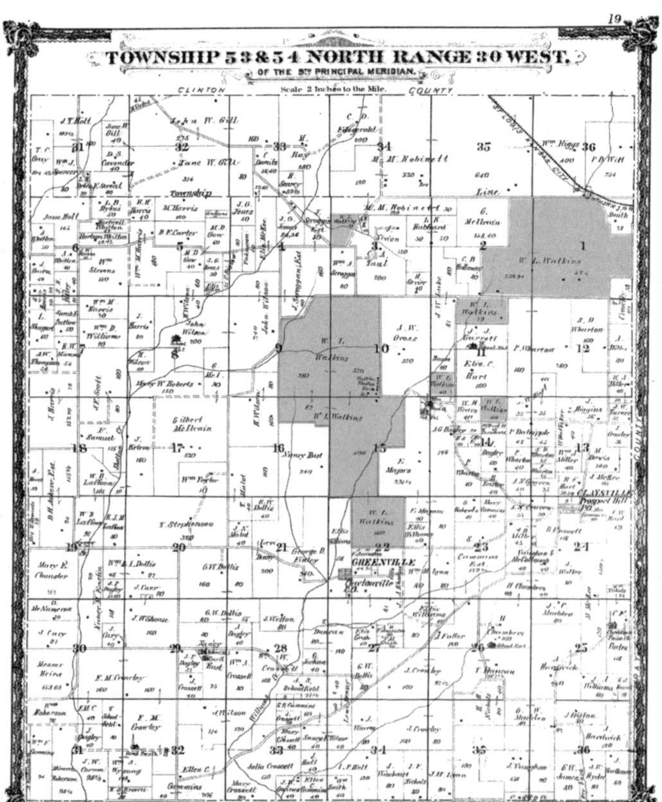

Section map showing the Watkins's holdings in gray. From the *Illustrated Historical Atlas of Clay County, Missouri.*

Farm or Plantation?

Although Watkins designated his Washington Township property "Bethany Plantation" (probably due to the large size of his holdings, coupled with his Kentucky background), he usually referred to it as a farm. The wide array of ventures projected by Watkins does not fit the antebellum qualification for a plantation: the extensive production of a single major cash crop. Yet the term might be appropriate in its connotation of self-sufficiency.

In his studies of the antebellum Little Dixie culture that developed along the banks of the Missouri River, R. Douglas Hurt has charted the evolution of "planters" in Missouri. He defines such as "an individual who owned at least twenty slaves and a minimum of five hundred acres." By 1860 only 4 percent of the farmers could be classified in the planter class. (Clay County had seventeen.) Though his land holdings—700 improved acres in 1860, with unimproved lands vastly outnumbering improved—surpassed the planter criterion, Watkins does not fit the planter category because he retained only three slaves.[a]

[a]Hurt, "Planters and Slavery."

claim in the 1870 census that his real estate holdings were valued at $122,000.[4]

For Watkins and others at this time, land was not so much a symbol of wealth or a source of income as it was an indication of family independence and long-range security. In turn, cash was not seen primarily as a means to gain consumer goods, but as a way to add to landholdings. Extensive landholdings symbolized an independent livelihood for Watkins, an asset that lured other family to Clay County, a measure of security in his old age, and a legacy for his heirs.[5]

A Challenging Landscape

Watkins and his fellow pioneers may have been attracted to the region by its rich soil, abundant game, and cheap land prices, but the physical landscape presented them with a number of challenges to settlement and prosperous competition. Oral histories tell of shoulder-high prairie grass, sky-darkening flocks of wild fowl, and ever present rattlesnakes. A modern historian forthrightly proclaims, "Much of the territory is very broken, rough and rocky and worthless for agricultural purposes. Many small streams, all of which ultimately run into Fishing River and its forks, head in the township. In many places picturesque bluffs are found along these streams, and the scenery is beautiful to look upon, but hardly appreciated by those owning the land." Nevertheless, Watkins apparently was not daunted by what he saw. Perhaps his discerning eye focused on latent assets such as the limestone outcroppings, which could be quarried for use as foundation stones or lintels for his structures, and the local clay, which would (after seven attempts to develop the correct concoction) provide the raw material for his brick kilns.[6]

Watkins evidently selected his plots because they were margins between prairie and woodlands. He was too late in the area to buy the lands noted as "first rate" by surveyors—or "land lookers"—who most probably relied on the tradition that a soil's fertility was revealed by the quantity and quality of its flora. (Large and dense hardwoods would be supported by "1st rate Soil," a profusion of less noble trees indicated "2nd rate Timber," and "barrens" supported nothing but grasses.) Sections 9 and 10, where Watkins would center his enterprises and build his homestead, were described thusly: "Land this mile gently Roling 2nd rate soil Prairie land gently = 2nd rate soil." Abutting the open lands were timber stands of oak, walnut, hickory, and elm. The understory of "hazle" indicated the wooded acres were not thick enough to block sunlight. As an aspiring stockman, Watkins undoubtedly saw this mixed environment as suitable for raising Berkshire hogs, Shorthorn cattle, and especially mules for export to the Lower South.[7]

Watkins and his neighbors spent many an hour clearing farmland from forested plats. This was an era when the average family farm would be lucky to clear 10 acres per year. Yet by 1850, historians estimate, American farmers had claimed 114 million acres from woodlands. Census takers in 1860 noted that Missouri had almost twice as much farmland (20 million acres) as in 1850. When Watkins acquired a steam engine for his factory in 1861, such deforestation of Washington Township would accelerate. Trees were obstacles for agricultural pursuits, but lumber and firewood were commodities for other endeavors, and the fruit trees and Osage orange that Waltus (and subsequent generations of Watkinses) planted in their place had value as well.[8]

FARMING FOR PROFIT

Expanded Markets

Waltus Watkins was not a lonely pioneer scrounging for subsistence, but rather an aspiring commercial farmer, whose endeavors would rely on numerous kin, neighbors, and hired help. R. Douglas Hurt has noted that fledgling farmers such as Watkins practiced subsistence or "safety first" agriculture initially but rapidly began sending their surplus production of diverse crops to market via the Missouri and Mississippi Rivers. "Subsistence agriculture was something to be endured until, having worked hard with ax, plow, and scythe and frequently slaves, they were able to clear the land, turn the soil and earn a profit." In moving beyond subsistence farming, Watkins made good use of available resources. His soil was more fertile than that in the East, and he estimated the yields per acre were on the average twice as great. He worked hard at increasing the yields from his fields, as well as venturing into processing those yields; oral histories from farmhands who had worked at Bethany, recorded at the beginning of the twentieth century, claim that throughout his career Watkins would purchase innovative implements to exploit the land.[9]

Peter H. Cousins has observed, "Among the many romantic myths about their past that Americans cherish, perhaps none is stronger than the belief in the independence and self-sufficiency of the family-owned farm and of the rural crossroads village community.... A more accurate way to view rural life at this time would be to portray the farm as an outpost—albeit a remote one—of a larger economy."[10] As such an outpost, Bethany was well situated. The placement of Watkins's holdings—within a day's journey to the county seat at Liberty, within twenty miles of landings on the Missouri River, and astride the emerging thoroughfare connecting Richmond and Plattsburg (with further termini at Lexington and Fort Leavenworth)—supported his agricultural enterprises. Being located adjacent to the state road gave Watkins a ready market. From the 1830s onward, military wagons ferried supplies from Liberty to St. Joseph, which was then the most active commercial center in the region. An irregular stagecoach service operated as early as 1842; a decade later five regular stage lines traveled the route. Watkins offered trade in farm goods, ground meal and flour, lumber, and later woolen processing and finished products, helping contribute to Clay County's standing as a market surpassing the nascent Kansas City.

As transportation improved, Watkins, like other farmers, could sell his surpluses through expanding market outlets. Through family connections in Kentucky and Tennessee, Watkins kept abreast of commodity prices, from beeswax to slaves. These agents, in turn, completed sales of goods he shipped by river and packhorse. By 1836 he was also in the freight business, contracting for his wagons to carry supplies of forage and lumber west to Fort Leavenworth.

Watkins regularly corresponded about crop prices and conditions with his father-in-law, Spencer Holloway of Jessamine County, Kentucky, who served as one of his role models as a farmer. (The younger man ultimately proved more successful, so that in 1842, Spencer confided, "But when I think of Mr. Watkins having a thousand barrels of corn by him I begin to think that I am a poor farmer.") The two discussed shared agricultural interests such as hemp, a crop that had migrated with Kentuckians to Missouri. Missouri farmers strongly favored the dew-rotted hemp used for cordage and bagging in the cotton kingdom of the deep South, and averaged 700 pounds per acre. By 1852, Clay County hemp growers were contributing one-third of Missouri's total hemp production, and on the eve of the Civil War, three manufacturers in the county were processing 900 tons per year into rope and twine. However, the market was volatile. Kentucky producers seemingly controlled prices; international events such as the Crimean War affected the often superior European product; Missouri growers were handicapped by reputations for growing a product of only fair quality and for handling it poorly; the federal government never favored Missouri efforts to guarantee a market through U.S. Navy contracts; and the panic of 1857 caused price contractions. Unless the price was predicted to reach $100 per ton, Watkins and other Missouri farmers held back from planting hemp. In 1844, Holloway noted, "We have been again blest with fine crops of corn and hemp though nothing commands a price.... So you see everything is moderately low here." The same year, Watkins was plagued by rains that produced prodigious flooding: "We have had the most distressing Season for the past two months that I have ever Saw in my life...it has ruined the hemp nearly ruined the wheat and it is the most gloomy prospect for corn you or I ever saw." Two years later, Watkins did not bother sowing hemp. Perhaps he was deterred by the risk, or by hemp's labor intensiveness throughout the calendar year (which led the region to rely on slave labor). At their peak, Watkins's farms had over a thousand acres planted in cash crops, primarily in corn, oats, and hay. In addition, during the 1860s and 1870s, bluegrass and timothy were planted to improve the pasture land.[11]

Livestock Breeding

Although Watkins gained a reputation as a commercial farmer, he was held in even higher esteem as a stockman. In the 1830s, he purchased a registered Spanish Jack, aiming to raise high-quality Missouri mules. For decades he worked in partnership with John Handy, who evidently made annual trips west to drive stock back home to Kentucky. In 1846, Handy and a partner drove 130 mules and 2 horses from mid-Missouri to central Kentucky (a twenty-three-day trip). He was none too impressed and claimed that "the stock in Mo are too small and badly kept for to be as profitable as they would be if they were better kept, a few large Jacks would prove of great service to your Country." He proposed to Watkins that they try a joint venture for 1848–49. "I know of nothing that a farmer could do in Mo that would be as profitable as raising mules. I believe you could make at least $1,000 a year on your farm.... it is a business that requires good Judgment and attention as well as industry to succeed well." The key was for Watkins to buy a 4-year-old jack "of good Size and a good breeder" in Kentucky for service in Missouri. Handy in turn would be responsible for getting at least fifty mules to the Kentucky market for sale each spring. The timing of the scheme proved propitious in light of the demand created by the California gold rush. Correspondence indicates that the Watkins–Handy mule trade was extant in 1852, when Handy had Watkins buy fifty-four head at $45–$50 apiece, hoping to sell them in Kentucky at $65. Their market also extended to the East Coast; in 1865 Waltus's son John helped one of his Kentucky cousins deliver mules to Washington, D.C.[12]

While the census records for 1850–80 show Watkins owning between five and thirty mules, he actually sold far more. It is believed that as many as 150 mules were often to be found grazing at Bethany. The Watkinses were primarily known for their large Spanish Jacks; other quality stock came from Shaker sources touted by Handy. Watkins maintained both a jack and a stallion license; in addition to breeding his own horses (Prince and Alhambra) and mules (Montana and Napoleon), stud services were available to the public for a fee. He also possessed an uncommonly large number of oxen. Some were used to power his hemp press, sawmill, and gristmill; others might have been harnessed to freight wagons, used in sod breaking, or sold to pioneers in their trek to the far west.[13]

Moving Into Capitalism

By the late 1850s, Watkins and his contemporaries knew that their farm produce could purchase three times as many eastern goods as had been possible in the previous generation. Transportation outlets, market rates, and increased cash income all stimulated commercial farming at Bethany. Watkins's entrepreneurial abilities were challenged by factors such as capital, labor, mechanization, and market fluctuations. He did not succeed in every endeavor. Early efforts to establish cotton milling operations in Liberty faltered; later efforts at raising broomcorn at Bethany proved to be a short-lived fad; investments in neighboring Ray County coal mines later ended in the company's bankruptcy. Still, his willingness to make adjustments as he advanced in the market economy, plus his persistence, led to rising prosperity. Self-sufficiency grew into a comfortable existence for the Watkins family. As Waltus's wealth increased, he was more able to establish his children with assets of their own. In sum, his view broadened to include cash and tangible property as well as land as indicators of his family's security and social position.

In the 1840s, Waltus Watkins had invested in many ventures heralding his increasing commercialization and capitalism. Analyst Walter Licht has written that "throughout the nineteenth century,...American farmers vastly expanded the agricultural base of the nation and that agriculture continued to represent an essential building block of the economy.... America became an unfettered market society."[14] Two examples illustrate Watkins's participation in these processes.

Bethany Gristmill

It is not clear when Watkins commenced milling, but his gristmill operation at Bethany was established at least by 1849. In March of that year, his agent in St. Louis wrote that he had visited with a supplier in search of $125 worth of machinery: "They have altered the pattern & Improved it they say and can furnish you a much better article than the one you got before." The supplier, the Todd Company, emphasized

Grain mill, view looking from the southwest. Photo by Jet Lowe, 1978 (81).

that "corn should never be ground on stones that you use for wheat it is too hard and will get the stones out of order if there is the smallest defect or difference in the different blocks." Watkins hired George Ferhand and Grundy Grimes as the millwrights to set up his gristmill. In May, Elizabeth Carter observed during a visit to Waltus and Mary Ann, "They have had a long and tedious time of it, with the mill and post office both. But they've got the mill running now and I think it will be profitable, and the post office has now been moved from here. Waltus has a good miller." Although there were rivals (the industrial census of 1850 showed four other gristmills then operating in Clay County) his operation gained a superior reputation. In 1875, Watkins upgraded to French buhrstones for both corn and wheat operations. These moderate-sized stones, weighing between 1,400 and 2,000 pounds per set, yielded a whiter flour than other stones and were more durable, lasting longer between dressings (leveling and adjusting).[15]

Gristmills symbolized the transition from subsistence to market economies. Between 1810 and 1860, the total value of flour produced in American mills grew from $14 million to roughly $250 million, largely due to new mechanized and labor-saving equipment. In 1840, Missouri had 690 gristmills. Because transportation links in that era were largely undeveloped, the mills usually served a market within a twenty-mile radius, creating a network from which a town might later evolve. The miller was often the first mechanical engineer in an area, and his mill commonly became the hub for associated industries such as blacksmithing. "The gristmill was a place where mechanical skills could be learned and refined, and where a miller unusually adept with mill machinery could expect to earn more than competitors.... The operation of a gristmill involved a wide range of social interactions that constantly tested the miller's abilities." Like other gristmill owners, Watkins gained further recognition among the wider community that sought his services.[16]

Mills also provided social interactions for members of emerging pioneer communities. For example, some of Watkins's more distant customers stayed overnight with him at Bethany while their wheat or corn was milled. Farmer J. A. Nance's diary for 1855 recorded four trips to "Wadkins Mill," each time returning with five bushels of cornmeal. An anonymous author's reminiscence of boyhood recollected "when I would go to the mill with a sack of corn on a horse, when the trampwheel mill was run by ox-power." There the boys "would engage in parching corn, running, wrestling, etc."[17]

Originally the gristmill was located northeast of the cabin on the Lane. Probably in 1868, the apparatus was moved inside the factory to take advantage of steam power and enhanced mechanization. As woolen production crested, Watkins was able to allocate profits to modernize the gristmill operations, for example purchasing a new wheat stone in 1875. This portable model from the Todd Company was housed in a freestanding frame rather than incorporated into the structure of the building. Judson wrote Mattie about this development, "I suppose you have heard all about the new mill. We have been running it some, half dressed, and it makes very nice flour now, so by the time it is dressed right nicely it will be quite a success I think... [William Reed] is the happiest miller extant. The new buhrs are the nicest you ever heard of and as hard as flint."[18]

The wheat grinding process produced four grades of meal; XXXX flour was premium, while X was very much like flours sold today as "stone-ground." The remaining bran and coarse particles were known as "middlings" (or shorts) and were used for animal feed. Corn meal was fairly uniform. Customers specified what portion of their flour they wanted in each grade and gave the mill a percentage of the yield as payment. In the early years, Watkins gained perhaps one-sixth or one-eighth of the meal as his fee for processing. In the years 1868–98, miller Reed was, according to some ledgers, paid one-third of the gristmill sales (as well as his wages). In later years, local farmers constituted only about half of the mill's business; the other half of the mill's grinding was bagged for sale on consignment through stores in the region.[19]

The gristmill was a profitable venture for many years. In fact, the miller and his assistants often provided a greater margin of profit to the Watkinses than did woolen manufacturing. In the industrial census of 1850, Watkins reported an investment of $1,500 for two stones attended by one hand (paid $16 per month) and powered by horses.

Twelve thousand pounds of wheat and corn had been ground by the buhrs. He estimated that his mill added $1,000 value in the processing of these crops. By the time the 1870 census was taken, he was able to report $10,000 in gristmill operations, with $4,000 in profit; Reed and two hands had worked six months, converting $7,000 worth of grain into $12,700 of sales. In 1880 the gristmill was capitalized at an estimated $1,000. The miller took in 10,000 bushels of wheat and an equal amount of other grains, valued at $15,000. Two employees earning $575 in wages (six months full time and four months half time at $1.75 and $1.00 per day) produced $17,000 worth of flour (2,000 barrels), cornmeal (540,000 pounds), and feed (160,000 pounds). Profit here was estimated at $1,325.[20]

By the turn of the century, large roller mills in nearby Kansas City and in Minnesota proved too competitive for small regional enterprises like Watkins Mill. When the Watkins steam engine broke down while grinding corn for feed in 1904, John closed down the gristmill operations for good.

Trading Outside Clay County

A second illustration of Watkins's early participation in the market economy involves transactions undertaken on his behalf by Anthony Minter in 1843. Described by a relative as "a fine old gentleman but a singular genius," Minter restlessly traveled the family circuit between Kentucky and Missouri. In the summer of 1843, Watkins sent him on an extensive trip to sell farm produce in St. Louis and New Orleans, after which he proceeded to Chicago and New York for purchases.[21] In New Orleans, Minter hawked casks of beef and ham, plus lard, hides, tobacco, and hemp. In correspondence with Watkins, he conveyed buyers' complaints that poor packing had made the meat less than prime. In addition, he had had to leave some meat to be sold by an agent in St. Louis, and at least $400 of his sales would be covered by a note from a New York bank. Defensively, he concluded, "If you should take any exception to my actions, I shall be able to explain all the motions that prompted me." A month later, Minter sold hides, tobacco, and hemp in New York, operating through the house of Thompson and Adams. Of his purchases there, he reported, "We got some good Bargains & some not so good at least I shall not loose [*sic*] on any thing I have." Mary Ann would receive "a First Rate Blue Cloth Cloak," the physicians in Washington Township would replenish their botanical supplies, and the Franklin Library would gain $46 worth of new books. Perhaps feeling that Minter had not gotten top prices for their goods, Waltus chided his uncle for being overly prudent, creating hard feelings that lasted throughout the winter. Nonetheless Minter resumed a similar role on his nephew's behalf in 1844 in St. Louis and in Lexington, Kentucky.[22]

CENSUSES OF 1830–50

Missouri's first state census was taken in 1830. As Watkins spent that year in transit, from Kentucky to Liberty and back, he was not counted as a Missouri inhabitant. Ten years later, both Waltus and James Watkins were listed as male heads of Missouri households. Within Waltus's unit were thirteen folk. The 1840 census taker was not then required to list names; rather, he recorded one male under age 5, two between 20 and 30, and two more in the 30–40 cohort. These would have included Waltus and his young son George; the others could have been kin or hired hands. Two women were listed in the 15–20 age bracket, two 20–30, and one 40–50. These probably included Mary Ann, Waltus's mother Jane, and his sister Rebecca. The census taker noted that six of these persons worked in agriculture, most probably the four adult males plus some slaves in Watkins's possession. These were described as one boy under 10, one male 10–24, and one female whose age could have been in the 5–10 or 10–24 category. (The counter's mark is ambiguous.) This was the household that first established the farm at Bethany.[23]

The 1850 national census was designed to be more detailed and more personal. The Watkins family (number 849 in Washington Township) was headed by Waltus, age 44, a farmer whose estimated wealth was $15,000 and whose native state was Kentucky. Mary Ann and their five offspring were also enumerated, and it was noted that three of the children had attended school within the year. Jane Watkins (age 69) and Absolum Dagley (22), a white male laborer born in Virginia, were listed as part of the household. Curiously, no slaves were recorded at the premises. The census of adjacent Platte Township revealed that W. L. Watkins owned a 33-year-old black female, a 19-year-old mulatto male, and a 2-year-old mulatto male. It is not known whether any of this trio was found in the 1840 list.[24]

The 1850 census included a separate set of questions on agricultural production. Watkins's responses indicated he was markedly above average for the region in size of holdings, diversity of crops, and their yields. Bethany was on its way to being a successful agricultural enterprise. Watkins estimated the cash value of his lands at $1,160; 650 acres were improved and 590 unimproved. A value of $3,945 was submitted for the livestock. He claimed his croplands that year had yielded 400 bushels of wheat, 600 of oats, and 4,000 of Indian corn. From the garden, only 10 bushels each of potatoes and sweet potatoes were noted. Processing of livestock included $645 from slaughtering, 120 pounds of wool, 300 pounds of butter, and 100 pounds of cheese. As the bulk of these figures were rounded off, the data probably represent Watkins's estimates rather than an exact count. (It is also likely that Watkins, to evade heavier taxes, underestimated his production. For example, the yield of fleeces, 2.4 pounds per sheep, is 50 percent below the norm.) His commercial endeavors were marked by $50 from the orchards, 250 pounds of beeswax and honey, 3 tons of hemp, and 20 tons of hay. A value of $50 was affixed to homemade manufactures. Omitted from the records were the yields of Watkins's sawmill and brick kiln. Still, the census records clearly show that in eleven years at Bethany, the hardy Watkinses had firmly established themselves.[25]

Historian Dan Feller has noted the need to revise our images of rapacious land speculators and virtuous settlers in the nineteenth century: "[C]ontrary to contemporary myth, the characteristic speculator was a doer, a booster. He stood to gain by promoting Western growth, not by thwarting it. Passivity was not in his nature." Waltus Watkins shared with his generation a labor theory of value. As he and his neighbors cleared land, planted crops, maintained roads, and erected institutions, their labors enhanced the value of their holdings. They also created communities, whose leadership niches men like Watkins were only too ready to fill.[26]

NOTES FOR CHAPTER 3

[1]James B. Bowlin, quoted in Barney, *Passage of the Republic,* 18; *Portrait and Biographical Record,* 216.

[2]Rohrbaugh, *Land Office Business,* passim. Watkins did not take advantage of the Preemption Act to acquire land, but he would indirectly profit when he provided services to migrants attracted by this liberalized policy. Land transactions in Clay County may be pursued in K. G. Bushman, *Index of the First Platbook.* Researchers in this topic as well as innumerable other facets of Clay County's heritage are indebted to the late Vera H. and William Eldridge, who provided the index to this volume and invaluable assistance to the authors.

[3]Prices are for 1839–70, the years for which documentation exists. Before the Civil War, Watkins was paying less than $10 per acre for farmland; purchases after the war were mostly in the $15–$20 per acre range. Most of Watkins's purchases in Washington Township were from prior investors. The first American settled there in 1822, and acquisition of public lands in the area sought by Watkins dated back at least to 1827, according to a review of platbooks of the General Land Office. In some cases, he acquired land by paying the back taxes of friends. Only 277 acres were acquired from either state or national lands, 40 acres at $1.25 per acre from the state in 1849, and 237 acres at the same price from the national domain in 1854. Notes of the First Platbook of Clay County, Missouri, 1819–1875, WMSHS.

[4]In 1834, after his brother James sent money with instructions to invest it "in the way you think it will be most profitable to me," Waltus bought 360 acres in Washington Township. It was his first purchase in the township that was to become his base of operations, though he did not take possession of the parcel until 1846, when James transferred ownership to him for $1,800. In December 1838, as he anticipated moving from milling to farming and more family arrived from Kentucky, he secured 99 acres (at $13 per acre) southeast of Liberty. Then in March 1839, he secured 80 acres for $300 in Section 15 of Clay County, and 40 acres in Section 35. On 1 April, he acquired, through a grant, close to 360 acres in Sections 9 and 10, adjacent to his holdings in Section 15. The ultimate plot for the Watkins homestead was part of 400 acres "more or less," purchased 1 June 1842, for $3,008. Watkins's land purchases in Washington Township might have been financed by the sale of his lot and home in Liberty; reportedly he was to receive $1,000 each in July of 1840, 1841, and 1842 from the purchaser.

After 1842, the bulk of Watkins's land acquisitions were to the east, near the Clay-Ray county boundary. This included several smaller parcels in Sections 2, 11, and 14. In 1848–49 and again late in 1853, nearly three sections, which now comprise the town of

Lawson (Ray County) were acquired. In 1872 he added 600 acres east of Lawson. James M. Watkins to Waltus L. Watkins, 30 August 1834, WMSHS; "Waltus Watkins Land Transactions," Research Files, WMSHS; Booker H. Rucker, "[Waltus Watkins] Land Transactions," Research Files, WMSHS; Agricultural Schedule, Tenth U.S. Census, 1880.

[5]In September 1882, near the end of Waltus Watkins' life, he and Mary Ann conveyed 2,238 acres in Washington Township and additional plots in Ray County to sons John, Joe, and Judson. "Listing of Debts assumed by John, Judson, and Joe [Watkins]," c. 1882, WMSHS; Agricultural Schedule, Tenth U.S. Census, 1880.

[6]Roney, *From Entry Fee,* 19. A modern geologic survey done when the state of Missouri acquired the Watkins Mill site noted, "The Watkins Mill area is typical of the loess-covered hills of the glaciated plain of northwestern Missouri.... The soil mantle is Marshall silt loam, a dark, rich, productive prairie soil" (WMSHS). The limestone found here is known to geologists as Bethany Falls limestone, named after Bethany, Missouri, the town that later emerged north of the Watkins holdings. Raymond Coveney, University of Missouri-Kansas City, private communication; see also http://www.kcwildlands.org/naturalwonder4.htm (accessed January 2004).

[7]Faragher, *Sugar Creek,* 40-43; General Land Office Survey, 1819 and 1861, Missouri Department of Natural Resources, Geology and Land Survey Office, Rolla, MO, WMSHS. Walter A. Schroeder, a close student of Missouri settlement patterns, has explained that French, English, American, and German settlers in Missouri used confusing definitions of *prairie* but that the term "came to mean a whole ecosystem, with its special wildlife, insects, soils and even its own climate. It had real and imagined advantages and hazards for human life and occupation." Schroeder, *Presettlement Prairie,* 2.

[8]Barney, *Passage of the Republic,* 19; McCandless, *History of Missouri,* 41-43.

[9]Hurt, *Agriculture and Slavery,* 7; Waltus L. Watkins to [?], [1840s?], WMSHS; Matthews (Sligar), Interview by Needham.

[10]Cousins, "Tall Timber," 7.

[11]Spencer Holloway to Waltus and Mary Ann Watkins, 23 May 1842 and 24 October 1844; Waltus Watkins to John William Luke, 2 July 1844, Waltus to Mary Handy, 12 September 1846, WMSHS; Hurt, *Agriculture and Slavery,* 108-24.

[12]John G. Handy to Waltus L. Watkins, 23 May 1847, Mary Watkins Handy to Waltus L. Watkins, 5 July 1852. As early as 1830, Missouri breeders were touting the advantages of their mules over horses. Lemmer, "Agricultural Improvement," 40-41.

[13]"Prince and Jack," Season of 1873, 1874 [studbook], WMSHS; P. Clark, "Breeding Practices"; Agricultural Schedules, Seventh and Eighth U.S. Censuses, 1850, 1860; Klinkenborg, "If It Weren't for the Ox."

[14]Licht, *Industrializing America,* xvi- xvii.

[15]Wyman Pittman to Waltus L. Watkins, 26 March 1849, WMSHS; Elizabeth Watkins Carter to Jane Minter Watkins, May 1849, WMSHS; Industrial Schedule, Seventh U.S. Census, 1850.

[16]P. A. Evans, "Merchant Gristmills," 318-21; Cowan, *Social History,* 72; Gordon and Malone, *Texture of Industry,* 240-41.

[17]Wyman Pittman to Waltus L. Watkins, 26 March 1849, WMSHS; J. A. Nance, diary, 1855, WMSHS; "A 1917 Historical Review" (author unknown), WMSHS.

[18]Judson Watkins to Mattie Watkins, 15 May 1875, WMSHS.

[19]"1917 Historical Review," WMSHS; Industrial Schedule, Seventh U.S. Census, 1850.

[20]Industrial Schedules, Seventh, Ninth, and Tenth U.S. Censuses, 1850, 1870, 1880; *Illustrated Historical Atlas,* 10.

[21]John R. Hardin to Waltus L. Watkins, 3 December 1846, WMSHS; Roney, "Old Watkins Mill," 61-62. New Orleans merchants urged Missouri farmers to sell their produce in the autumn, as farmers in the more immediate market of the Crescent City generally sold in the spring. Hurt, *Agriculture and Slavery,* 67.

[22]Anthony Minter to Waltus L. Watkins, 4 September1843, 4 October 1843, and 28 June 1844, Waltus Watkins to John Luke, 2 July 1844, WMSHS.

[23]Population Schedule, Sixth U.S. Census, 1840; Slave Schedule, Sixth U.S. Census, 1840.

[24]Population Schedule, Seventh U.S. Census, 1850; U.S. Census Bureau, Slave Schedule, Seventh U.S. Census, Platte County, MO, 1850.

[25]Agricultural Schedule, Seventh U.S. Census, 1850.

[26]Feller, *Public Lands,* 30; Young, "Congress Looks West," 79.

4

Leaders of the Neighborhood

As he shaped the elements that made up his personal realm—family, home, lands, businesses—so too did Waltus Watkins strive to shape the larger community in which he lived. A primary builder of a sense of community in the region, he helped establish a number of institutions and infrastructures for common use. However, his interests lay deeper than mere buildings and roads. Historian John Mack Faragher has observed, "The word *community* carries amid its many meanings a vision of men and women living in 'communion,' sharing something more than the rather prosaic bonds of neighboring, borrowing, collectively working and intermarrying." Watkins's civic activities were largely aimed at influencing that "something"—the goals and values by which the members of his community would live and conduct their affairs. As a community leader who espoused social harmony and a strict moral code, he championed restraints on individual behavior. He believed that, through family values instilled at an early age and discipline inculcated in schools, churches, and other institutions, individuals could be molded to accept their place in American communities. His concerns about individual behavior were perhaps fueled by the cultural changes brewing across the country. New norms, new beliefs, and new visions abounded in the 1830s. Both historians and contemporaries describe the era as marked by egalitarianism. The common man, not just the property holder, could now vote. Credit was within reach of more than just the wealthy. Many people openly displayed their desire to seize every opportunity for material self-advancement. Watkins and other community leaders feared that if everyone pursued their own happiness, the pillars of republican society would be shaken. Thus, in the reforms they advocated and the institutions they created—schools and churches in particular—they sought to convert, control, and uplift their fellows.[1]

Religious and Cultural Institutions

Mount Vernon Missionary Baptist Church

Initially the Watkins family worshipped at the New Hope Baptist Church, which Waltus had joined following his 1849 conversion experience. Then in 1857 Watkins determined to found a church close to his residence (New Hope being a distance of some five miles away) with the Reverend William Barrett. Waltus, Mary Ann, and 13-year-old Kate were among the first six families attending the Church at Mount Vernon (as it was originally called), along with the Andersons and Hollingsworths, who worked Watkins's lands as tenants. For about a dozen years, the congregation met in a small wood-frame church. Then in December 1869 that building burned down; only the pulpit was saved. Services were moved temporarily to a private residence and then to Greenville, pending construction of a new church building.[2]

Watkins's response to the destruction of the church was a measure of his commitment and also of his power in the congregation. It was he who determined that the new Mount Vernon church would be located at a site one and a half miles from the old, where it would be adjacent to the school, more convenient to his home, and closer to the state road, on land that he and Mary Ann donated for the purpose. Watkins and his stepnephew, John W. Luke, who was then the preacher, determined the size (seating for 150) and style (Italianate) of the new church. Watkins oversaw construction, donated the brick from his kilns—possibly two years' production—and provided limestone for the foundation. When the congregation was unable to cover building costs, he came up with the needed cash, reportedly as much as $1,200.[3] The church's obituary for Watkins would offer testimony to his contributions: "When this house in which we worship was being erected he left all things else and gave his earnest devotions to it until it was ready for occupation. He gave liberally of his means in support of missions & churches at home & abroad."[4]

The Mount Vernon minute book is replete with notes of Waltus's benevolence: firewood, oil for lamps, partial subsidy of ministerial salaries, and even $10 for spittoons. When subscription lists were made to cover church debts, the family led the list; one such roll (undated) for $211 from eighty-seven members puts Waltus and Mary Ann at the top with $11 each, while Carrie and Joe were allotted $4. Once Watkins specifically requested, over objections from fellow members, that both his Ray County and Clay County landholdings be considered in figuring his subscription assessment.[5]

Church records for the post–Civil War period repeatedly list Watkins in the most influential positions in the congregation. He is variously recorded as deacon, elder, and treasurer; served as clerk from 1867 to 1872; and annually was designated to attend sessions of the North Liberty Association. Watkins's influence in shaping the religious community at Mount Vernon took a number of forms, including drafting and subsequently revising the rules of decorum, and leading in the selection of ministers. After 1880, the minutes fail to mention Watkins's activity, leading to the inference that ill health forced him to withdraw from church affairs.[6]

Other Watkins family members also gave active support to the church. Daughters Kate, Carrie, and Lizzie (the last two had, by 1865, joined their parents and sister as members), together with Waltus, were listed among Sunday school teachers in 1870. Jewell, Judson, and Joe at one time or another were accepted as members; while the first two ultimately did not live up to church standards, Joe succeeded his father as a church pillar, serving in the 1880s as clerk and Sunday school superintendent. It is also probable that the family's cook was a member, as an undated membership list from around the time of the Civil War includes the entry "Eliza, a colored woman."

The congregation's rules of decorum anticipated fissures within the parish and failings on the part of individual parishioners. Frequently committees were appointed to counsel with the sinner and get him or her to confess, repent, and gain readmission. The Watkins family experienced such congregational discipline and forgiveness on a number of occasions. At some point, Carrie left or was removed from the church; when she was granted readmission in 1879, she divulged "doubt as to her conversion at the time of profession" but said she "had experienced real conversion" during attendance at the Fourth Baptist Church of Saint Louis with Joe Watkins. Waltus was cited in May 1872 for hauling hay on Sunday but successfully argued that doing so had been necessary to save his stock. The charge may have arisen from his conflict with a fellow member, William Dagley, who had previously had been charged with card playing. He in turn accused Waltus of violating the Sabbath. Later Dagley charged minister John Luke with pocketing all of the $5 from a funeral preaching. These episodes and many other charges traded among members forced the church to require written specifications of charges rather than rumors.[7]

Jewell in particular seems to have had trouble meeting the church's expectations, to his family's undoubted embarrassment. In January 1872, when he was not quite 20, Jewell publicly acknowledged he had acted "imprudently and recklessly." The church forgave him upon his confession; his father simultaneously resigned as church clerk. A little more than a year later, Jewell was one of three charged with intemperance. While his cohorts faced exclusion, Jewell was once again forgiven. In January 1874, Jewell "stated he was unworthy to be a member of the Church and asked to be excluded." Almost exactly a year later, he was restored, after acknowledging he had gone astray and expressing "deep Sorrow and penitence for his sins." Later records indicate he was withdrawn from membership due to lack of attendance in 1878 and again restored in 1886 (along with his brother Judson).[8]

Franklin Academy

"A self-sufficient Jeffersonian yeoman farmer might manage without reading or arithmetical skills," notes analyst Rogers M. Smith, "but Jacksonian commercial farmers, who purchased and sold goods via written agreements and the exchange of money and sometimes travelled to distant markets could not." Education, such as that available from the 1840s onward at Franklin Academy, also aimed at perfecting society.[9]

Franklin Academy was organized possibly as early as 1836. Because the state authorized but did not fund public education, the Academy was set up as a subscription school. Subscriptions were based on each family's economic circumstances; as the largest landholders in the district, the Robinetts and the Watkinses led the subscription drive. The Franklin School District dates to 1842; Watkins became a board member in 1844 and later served as inspector, scrutinizing the academic and personal qualificiations of potential teachers in the five-month term each year.[10]

In naming their new institution for Benjamin Franklin, the founders gave a clear signal of both the bourgeois and republican missions of the school. The school's functions were to instill model behavior among its students (because individuals had to be reformed before they could participate meaningfully in society) and to provide more advanced studies. To adjust adequately to the times, teachers demanded students master the 4 Rs—reading, writing, arithmetic, and rhetoric—so they could better prepare for work within the economy. Moral character and virtue, so necessary for responsible citizens in the American republic, could be gained from rhetoric, history, and other muses. In content as well as approach, the emphasis of the schoolmasters was self-control. The "McGuffey Readers" (W. H. McGuffey's *Eclectic Reader* series, which students had to purchase) emphasized the material rewards for virtue and unremitting labor; the texts set a high value on private property and taught that progress was best measured by growth in material possessions.

The academy was initially housed in a log structure. Perhaps to satisfy a restless teacher, Watkins spearheaded a drive that climaxed in 1856 with the erection of an octagonal brick structure, relocated closer to his home on the state road. As the largest landholder in the area, Watkins was assessed the most for this construction, which he also supervised and for which his kilns supplied bricks. The octagonal shape stemmed from a search by Watkins and fellow school trustees for the perfect school design. At that time the octagon was believed to be the ideal form, a fad originating with the well-known phrenologist Orson Fowler, who was also an amateur architect. Fowler claimed that for pedagogical as well as aesthetic reasons, the eight-sided shape would maximize learning, as the teacher could see and control the students better.[11]

One enrollment list, for early 1859, still exists. J. B. Elred, who boarded with the Watkinses, was then the teacher for forty-four students, eighteen of whom were females. Tuition for the whole student body for the five months totaled $261. Student ages ranged from 7 to 22 years old, with more students in the younger age brackets: twenty-nine were in the range 7–12 years old, ten were 13–18, and five were 19–22. (Older children tended to be more needed at home; some may also have mastered the desired subject matter.) The term ran from February 1 to July 1. Attendance varied from a high of one hundred days—a perfect attendance rate attained by nine students, including three of the five Watkinses enrolled—to a low of twenty. Males, drawn away by farm duties, were prone to less class time. There were five fundamental courses, plus six different mathematics offerings. Reading, spelling, and some form of arithmetic formed the core of enrollments; all but two students received instruction in reading and grammar, and two-thirds worked on writing assignments. There were no grades, as each student progressed at his or her own speed.[12]

Franklin Academy operated until 1870, when a public school was finally established in the same building. In 1878, the public school was moved to a new structure, and Watkins bought back the octagonal building at auction.

William Jewell College

William Jewell College in Liberty traces its roots to efforts by Watkins and his contemporaries to improve society in general and the Baptist ministry in particular. In 1843, after Dr. William Jewell of Columbia, Missouri, offered lands worth $10,000 to support the founding of a Baptist college, counties around the state competed to become home to the new institution. Clay County, with the Reverend Robert James and Waltus Watkins in the vanguard, mounted an endowment drive. When the goal of $60,000 was reached, Liberty gained the college, which held its first classes in 1850 with 137 enrollees.[13]

In 1847, Watkins was appointed as a subscription solicitor; three years later he assumed added obligations when co-solicitor Robert James died. In 1850, he began a twenty-year affiliation as a college trustee, which also included responsibilities as an examiner of the school's students. Financial problems forced William Jewell College to close in 1855, but renewed drives succeeded in securing sufficient funding to reopen in 1857. In connection with the 1857 drive, the *Liberty Tribune* listed Watkins as a key benefactor for the college. Three of the Watkins boys (George, Joe, and Judson) would attend the institution.[14]

Watkins's prominence in the county, his recent conversion to the Baptist faith, and his commitment to education combined to propel him in this civic quest. In Clay County, he found congenial allies in his educational mission. The moral economy and the political economy were interwoven. Many a citizen voiced agreement with the 1846 assertion by the *Tribune* that "it is an acknowledged fact that ignorance and republican liberty never join hands." From the establishment of public schools such as Franklin (second in the county) to higher education at

Clay Seminary for girls and William Jewell for boys, the county fathers provided facilities and opportunities for their youth to learn and prosper. In 1878, the *Liberty Tribune* offered a revised credo: "The future of Liberty in respect to education is brilliant. If she cannot be a busy mart she can be a great educational center." Watkins energetically sustained such hopes for a half century in Clay County.[15]

Debate and Library Societies

It was not only the youth of the county who learned and prospered through institutions created by their elders. Starting in the early 1840s, the Franklin Debate Society and the Franklin Library Association provided Watkins and other self-proclaimed "leaders of the neighborhood" (meaning the three adjoining counties of Clay, Ray, and Clinton) a means of reinforcing shared values while at the same time combating loneliness, isolation, and ignorance. Upwards of two dozen men met on Friday evenings from October through May in the Franklin Academy schoolhouse. Their orations were part of the national public ritual or pastime. For example, virtually every schoolboy in Watkins's generation read *The Columbian Orator,* which was first published in 1793 and ultimately sold two hundred thousand copies in twenty-three editions. Historian Robert Wiebe has remarked of this era, "America was a talking society where millions, not hundreds, shared the responsibilities for maintaining connections, spreading information, and persuading one another of the Truth."[16]

Watkins was at the nucleus of both the debate and the library movements in his community. In the Debate Society, he frequently served as president, furnished candles and firewood, and was the target of fines meted out both seriously and in jest to the group's leaders. In the Library Association, Watkins not only contributed a double subscription of $3 per year but also, in 1843, advanced half of the funds needed for the initial purchase of books. From the combined efforts of the Debate Society and Library Association came renewed support for education in the area, Watkins's favorite project.[17]

The Debate Society was founded in 1842; as true republicans they began by drafting a constitution, with an explicit agenda that was educational in a broad sense. In January 1843, at the insistence of Watkins and one of his close friends, Dr. James B. McCord, they passed the hat to sustain an affiliate, the Library Association. The library effort was deemed successful: "Considering the state of the currency and other circumstances quite a liberal contribution was made." This led to procurement of "the first two series of Harper's School District Library," for which Watkins advanced $25.31.[18]

The Society's minutes did not record the evidence, rhetoric, or logic of the debate teams and provided little note of the persuasiveness of the disputations. The secretaries, in their records, muted contention; they never commended nor condemned an individual debater in the minutes, nor commented upon teams' predominance or incompetence. The rotation of members through successive weekly debate teams emphasized consensus rather than conflict. Neither cliques of allies nor regularized sets of adversaries were apparent. Through the spring of 1844, the range and focus of topics were those confronting civilized people everywhere. Few queries ever posed a pro or con question. Rather than opposites, debaters argued options such as "Which has greater influence on mankind—wealth or knowledge?" A repeated topic was "Which is the greater stimulus to the actions of men—the hope of reward or the fear of punishment?" The debaters struggled with the norms of human behavior and the composition of communities.[19]

In 1844, having perhaps evolved a firmer sense of comradeship, Society members opened up a widening variety of queries to debate. As membership rolls expanded, more viewpoints would be heard in the schoolhouse. In the last fifteen sessions for which records exist, the query dictated an affirmative/negative dichotomous mode rather than dialogues. Now pronouncedly contemporary issues, charged with ideology, came to the fore. The presiding officer and audience heard arguments on the state school law, postage rates, and restrictions on immigration—all questions that elicited partisan stances. Ultimately the Society fragmented when members debated the question "Is it now practicable to form a general temperance society in this neighborhood whose members shall be bound in a pledge of honor to abstain not only from intoxicating liquors (as a beverage) but also from coffee and tea, costly apparel[,] in a word from every superfluidity not indispensable to comfort and happiness?" The controversy led, on 4 May 1844, to a cessation of the weekly meetings. The Society resumed meeting in the autumn of 1844, but enthusiasm flagged; members met less regularly, and attendance dwindled. Though the Society continued its library support, it had spent itself. By March 1845, it had apparently ceased to exist altogether.

The experiences of the Franklin Debate Society can be analyzed with models supplied by historians Don Doyle and Thomas Bender. The former claims that voluntary associations such as the Franklin Debate Society, "whatever their explicit mission,... performed very special covert roles by integrating community leaders, enhancing individual opportunity, safeguarding the middle-class family, and serving as schools that taught organizational skills and group discipline." In so doing, they "provided this mobile, diverse, and seemingly disorderly swirl of individuals with a cohesive and constant form of social order." Bender, focusing on territorial, emotional, experiential components of a community, states, "A community involves a limited number of people in a somewhat restricted social space or network held together by shared understandings and a sense of obligations." Equilibrium evolves between the "pursuit of individual opportunity and the need for community order." For a time, the Franklin Debate Society met

Within the Debate Society Watkins saw firsthand the pressures of conformity. When McCord, a botanic physician, was forced to resign from the Society because of his medical and religious beliefs, Watkins's courage failed, and he chose not to stand up for his friend. The doctor lost patients and was forced to relocate his practice to eastern Missouri. Perhaps troubled by his passivity during McCord's harassment, Watkins saved at least four letters, including one containing McCord's confession that he had lived "in torment" in Clay County. In 1846, when Watkins determined not to correspond further, McCord lamented, "Something has transpired which has caused a dampness or an abatement in your feelings of friendship.... I think you act through mistaken views." Watkins's own feelings have not been preserved in his manuscripts.[a]

[a]James B. McCord to Waltus L. Watkins, 10 April 1845, 29 August 1845, and 4 December 1846, WMSHS.

these dual purposes of individual and community. Yet the bonds of sentiment and self-interest were not durable. Membership was extremely fluid; participants joined when they perceived the Society as beneficial, submitted to in-group discipline and creed to varying degrees, then withdrew. When the Society disappeared, other institutions would carry on with community-building.[20]

Squire Watkins

Watkins played a number of official and unofficial roles in the civic affairs of his community throughout much of his adult life. The specific activities in which he chose to involve himself were no doubt influenced by his interests as a landowner, entrepreneur, and reformer, as well as his sense of duty.

His contributions to public culture in Liberty started soon after his arrival there and were documented in the early records of Clay's circuit court, where he appears as both a witness (during an 1833 murder trial of Elisha Cameron Jr.) and as a defendant (as a cosigner of promissory notes).[21] In 1836, he was one of five subscribers for a bond to sponsor the candidacy of John Baxter for sheriff. He also contracted to enclose the public square with a post-and-square rail fence, including four gates with stone steps to each. In 1838, he was appointed a justice of the peace for Liberty Township. Still, Watkins's participation in civic affairs did not raise him to the top tier of Liberty's prominent men. He was overshadowed by the town's original pioneers, such as John Peters and Shubael Allen; by fellow Kentucky migrants Alexander Doniphan and David R. Atchison, who garnered political laurels; and by Major John Dougherty, who parlayed service as Indian agent and fur trader into building the county's grandest mansion, "Multonomah," just west of the county seat.[22]

After moving to Washington Township, Watkins was again appointed justice of the peace. In this capacity, he personified law in his community. "Justices decided simple legal cases, performed marriages, and acted as notaries public. One of their number was appointed each year by the county court to divide roads in their township into districts and assign people to maintain them." According to his autobiography, not one of his verdicts in thirteen years of service was overturned by the circuit court. That, together with the fact that Watkins continued to retain this important community office, strongly suggests that his contemporaries found him a man of integrity in determining justice. His prominent status was recognized in the minutes of the Franklin Debate Society, where he (and only he) is referred to with the title "Squire."[23]

Watkins gained additional prominence and established a public service for his neighbors when his petition to be appointed postmaster was approved on 19 August 1844. His duties, handled from a desk in his home, included placing handwritten postmarks and dates on outgoing mail. His own mail, per custom, went free of charge, provided he placed his signature next to the postmark of Bethany. This postal service lasted only five years; when the town of Bethany in Harrison County successfully petitioned for the designation of "Bethany Post Office," Watkins chose to terminate his postal duties rather than seek a new name. Thereafter the Watkins mail was addressed to the post office at Prospect Hill in Ray County.[24]

Politically Watkins was a Democrat. Though never a candidate for elective office, he sought and gained power within the county political structure, which was usually controlled by Democrats. (Ironically Clay County did not pledge its loyalty to its namesake, Henry Clay of Kentucky, who played a key role nationally in leading the Whigs against the Democrats in the three decades before the Civil War.) Historians have discovered a "close and visible fit between wealth, Whiggery, evangelicalism, and reform." When Watkins first came of political age, he had yet to show any of these tendencies; in later life he manifested all of them, yet remained a steadfast Democrat.[25]

There is no direct evidence of Watkins's participation in the political forums, bonfires, barbeques, and parades that constituted the most visible cultural expression of antebellum America. (The Missouri manifestations of political activism, from stump speaking to vote canvassing, were immortalized in a number of 1850s paintings by George Caleb Bingham, himself an aspiring Whig.) His chosen profession was that of an entrepreneur, not a politician. He was not, however, uninvolved in the partisan political process. In 1852, he served as one of the twelve delegates from Clay County to the state convention.[26]

Railroads and Towns

The "neighborhood" in which Watkins moved included not only Washington Township, but also other sections of Clay County and townships to the east (Ray County) and to the north (Clinton County). In this area, as in other frontier regions, communities emerged as local residents were drawn together by shared commercial and other interests. Some endured and grew into towns; others dwindled or died out altogether. Often the determining factor was the settlement's location relative to key transportation outlets, especially railroads. For example, the significant towns that did arise in the 1870s–80s (Kearney, Lawson, and Excelsior Springs) were creations of the railroads that skirted the township. Every hamlet in Clay County that was not on a railroad eventually faced extinction.

Three towns near Watkins's holdings illustrate this rise and fall of commercial nodes along the Richmond–Plattsburg thoroughfare. Haynesville, just over the line in Clinton County, was platted in 1842; for a time it was home to gristmills and wool-carding operations.[27] It fell dormant after 1867, as rail lines were developed elsewhere. Greenville, just south of Bethany, was founded in the 1840s by Ellis Williams and would eventually include his hotel, two churches, educational institutions, and a post office (Claytonville). In 1850, its population was estimated at 150; the *Illustrated Historical Atlas of Clay County, Missouri,* published in 1877, showed thirty-eight planned lots on two streets. Yet in 1885 the town was disestablished. A third community, southeast of Watkins's homestead and abutting the Ray County line, was Claysville (Prospect Hill Post Office), which came to life in 1852 along a prominent stagecoach route. The town featured wagonmakers, blacksmiths, the Golden Bar saloon, and later a racetrack. By the time of its listing in the 1877 atlas, it had become the address mainly of farmers and stock dealers. Claysville's demise resulted from railroad beds being routed to the east and a depot established at present-day Lawson.[28]

Watkins recognized the potential importance of rail transport for the region's development. As early as 1847, he represented Clay County at regional conventions in efforts to have railroads routed through the county. In 1859, the Hannibal and Saint Joseph Railroad was completed through the northern tier of Missouri; in 1868, a railroad bed, later known as the Kansas City and Cameron, bisected Clay County on the north–south axis, with a second line added later parallel to the Missouri River. However, much of the county did not have ready access to the railroads. As a result, industrialization did not take hold, and commerce languished.[29]

Census of 1860

When O. P. Moss and other enumerators traveled the byways of Washington Township in 1860, they were undoubtedly struck by the economic growth and diversification of the area. Although farming was the predominant occupation, the township now boasted a miller, an engineer, two mechanics, three wagonmakers, five carpenters, five blacksmiths, and a seeming abundance of teachers and doctors. Watkins, still listing himself as a farmer, now estimated his wealth at $21,000 in realty and $15,000 in personal property. Only one other inhabitant of the township exceeded Watkins in estimated worth. The Watkins family now included nine children at home. George, at age 23, had no occupation listed; six of his siblings had been sent to school that year. No relatives, boarders, or slaves were in the household count.[30]

The companion agricultural census noted that under Watkins's management, many ventures had not changed in ten years, but some noticeable realignments had occurred. For example, he had added only 50 acres of improved lands since 1850 but had more than doubled unimproved acreage to a total of 1,400 acres. Livestock herds had been adjusted. While he had slaughtered $1,500 worth of animals, he estimated the value of his stock at only $2,000. He had thinned herds of mules, milk cows, and working oxen. Sheep, other cattle, and horses numbered about the same as in the last count. Croplands were noticeably less productive of corn, wheat, and oats. More honey had been produced, plus 200 gallons of molasses. Household manufacturing was estimated at $75. Butter production had increased, but cheese decreased. According to this picture provided by the census, Waltus Watkins had, on the eve of the Civil War, attained a comfortable plateau. Now, if he wished to gain added wealth, he needed to acquire more labor for his farms or more livestock for his herds. Alternatively, he could invest capital and energy in industrialism.[31]

The Eldest Son

The effects of Waltus's shaping influence can be seen clearly in his eldest surviving child, George Spencer Watkins. To provide him with a higher education befitting his class and prospects, the Watkinses sent George away to school, first to the newly opened William Jewell College for two years and then, when that institution closed in 1855, to Georgetown College in Kentucky. As he moved away from his home base, George discovered a very different world from Bethany. An acute and thoughtful observer, he filled his letters home with reports of widespread cultural differences and social changes in the years preceding the Civil War. The letters reveal a young man weighing these experiences against a clear set of family values and seeking to find his place in the world.

In late summer of 1855, 18-year-old George left for Kentucky, escorted by a Franklin School teacher. Passing through bustling St. Louis was an eye-opening experience: "Sounds like noise of hundreds of

George Watkins. DNR, WMSHS.

sawmills shelling machines carriages and drays the levy is crowded with boats for miles and the shore is thronged with people of all descriptions Never will have conceived what a larg [*sic*] city was." His destination lay within the family network; John Luke had already attended Georgetown, and the Handys would be nearby to look out for him. Initially George was positively impressed by his fellow students, whom he considered "some very fine young men...[a] very clever set." Later however came the description of his classmates, already quoted, as "weak puny and efeminate." His self-esteem, measured largely in athletic contests, grew.[32]

In the fall George attended a lecture "on the duties of men and the rights of women given by one of these hard mouthed northern women. She is the first lady I ever heard lecture." He was captivated. Though he did not agree with her stance, he was entranced by her independence. This woman stood in contrast to the "Ky flirts and coquettes" he encountered in visiting the Handys. George entertained his sister Kate with tales of fashionable Kentucky ladies. "They dress very gay and seem very proud and are no doubt as heartless as a brickbat." He particularly focused his sarcasm on hoopskirts, hilariously recounting how one fellow had been entrapped.[33] The girls at the Georgetown seminary bored him with their readings "of love and bright hopes of broken hearts and bright anticipation." Still, he defended the girls' right to be there; in the boys' debate society, he found fault with those who contended males should gain preference in education. George wrote to his parents, "I think neither should have the preference, but that the girl should have as good an education as the boy not that the girl should study the same things as the boy but that she should be as well prepared in her sphere as the boy is in his." Such a stance affirmed the Watkins plan. Later he opined, "The Ky boys [are] three times as wild" as their peers in the north, "and I feel that is owning to the climate. The southern people are quicker incited and are mor [*sic*] high tempered than the northerners."[34]

Sectional differences became clearer to George as he grew more venturesome. Without consulting his parents or his Kentucky elders, he determined to spend the 1856–57 academic year at Madison College, in the center of New York State. It was a bold personal statement. After the fact he told his parents, "I thought I would have advantages here that I would not have at george town and because it is cheaper and a person gains a knowledge of the world by traveling and mingling with those from different parts of the world which he cannot gain in any other way." In his further defense, he listed Missourians at Madison.

While he immersed himself in studies befitting a refined gentleman, George continued to offer critiques of the cultural differences he encountered. Northern preachers espoused political sentiments from their pulpits in the election year and, more startling, were "manufactured preachers," meaning they relied on prepared written sermons. The folk around bucolic Hamilton preferred manufacturing and dairying to the agricultural pursuits he was familiar with from western Missouri and Kentucky. He recounted the "great many curiosities" in a local museum and the captivating music of women's choirs accompanied by organs in the churches. New York food and dress was contrasted to the tastes of the proud Kentuckians he had left. Along the banks of the Ohio River and the Erie Canal, he had observed "a great many men in the different cities that I passed through some were going one way and others another but all seeking for riches and pleasures which they falsely call happiness." Such a portrayal of urban entrepreneurs might have been George's way of leading his parents to accept his current enthusiasm for the arts and languages and contemplations of a career as a teacher.[35]

By February 1857, George had been transformed. He announced, "I am really glad that I came to new york evry thing is so different from what it is in the south I would likely have never fully realized the difference even between the words north and south had I never lived in the north I can almost say the words and thoughts are different I dont think the people are so bold and enterprising as the western." He related that he had encountered Yankees—"not so good at repartee and satire and irony as I expected to find"—and Irish—"the wilyest people that I ever met."[36]

Perhaps due to lack of funds, Waltus called George home before his studies were completed or his conversion deepened. After two years' absence, he arrived home on 1 August 1857. Irony awaited him. Earlier George had congratulated his father on his sawmill, "a money coining machine to one that could stay right with it and keep it going but hirelings I do dispise [*sic*] for the manner they work and eat."[37] Watkins assigned the aspiring young man of refined tastes the role of sawmill supervisor. George rose to the challenge and succeeded.

The Approaching War

In the summer of 1856, as George Watkins pondered the implications of the impending sectional crisis for himself and his family, he wrote to his parents, "I fear Kansas will destroy the Union and hand it as a prey to the black Republicans."[38] The Kansas-Missouri border warfare of the 1850s and the intense battles of the Civil War certainly placed Watkins's domain in jeopardy. His anxieties over an expanding agricultural market, and transition from a rural agrarian economy to the more impersonal market system associated with industrialism, were compounded by threats of disorder and fears of disunion. Furthermore, changing social and economic philosophies brought him, and others, to the realization that slavery was "an unnatural, and unnecessary, system of unfree labor." The world that Watkins and his peers had worked so hard to build and shape seemed about to collapse.[39]

NOTES FOR CHAPTER 4

[1]Faragher, *Sugar Creek,* 170.

[2]The church fire occurred a year before a fire set at Franklin Academy (resulting in $4,000 in damages). In the latter episode, schoolmaster Albert Chase accused Lewis Turner, Lewis Sullenger, and John Ryle of arson, but the county court determined, due to insufficient evidence, that they were not guilty. The *Liberty Tribune* asserted that the Franklin fire was "believed to have been the work of some scoundrel running at large, defrauding the penitentiary." *History of Clay and Platte Counties,* 499; *Liberty Tribune,* 31 December 1868; Clay County Court Records, Book 17 (3 January 1871–May 1873), nos. 222, 226, 235, 380, and 612, Clay County Archives.

[3]Waltus Watkins obituary, from *History of Clay and Platte Counties,* 506; Mount Vernon minute books, 1865–87 vol., pp. 17–19, WMSHS.

[4]Mount Vernon minute book, 3 May 1884, WMSHS. Watkins's support of the church also enabled him to engage in the paternalism that he applied to the school, his mill employees, and other facets of community life (see chapter 8).

[5]Undated subscription list, Mount Vernon minute book, 1865–87 vol., WMSHS.

[6]Son Joe would try to fill the void, but by 1918 membership dwindled and the church disbanded. In compliance with the 1875 deed, the property was then returned to the Watkins's estate. John subsequently stored hay in the structure. Mount Vernon minute book, 1901–28 vol., WMSHS.

[7]In 1873, Dagley was excluded from church membership. Interestingly the church never reprimanded Watkins for operating his woolen factory on Sunday, which he did frequently. Mount Vernon minute book, 5 July 1879, pp. 65 and 87, WMSHS; Lentz and Wilson, "Mill Ran Today."

[8]Mount Vernon minute book, 5 April–11 July 1873, 25 January 1875, WMSHS.

[9]R. M. Smith, *Civic Ideals,* 216.

[10]"Franklin Academy," Research Files, WMSHS; DAR, *Centennial Souvenir;* Ethel Massie Withers, "Education in Clay County," Research Files, WMA Collection, WMSHS.

[11]In 1871 Watkins led a successful petition drive to reroute the Plattsburg-Richmond Road where it ran north-south by his mill before it turned east to John Luke's property, then south to the church and school. In the spring of 1873, the county court loaned him "what amount of money is on hand belonging to the different school townships" in Clay County. He was to have access to these funds at least to June 1 at an unspecified rate of interest. Clearly, Watkins's interest in education and religion was not altogether altruistic! Clay County Court Records, Book 17 (3 January 1871–May 1873), nos. 222, 226, 235, 380, and 612, Clay County Archives; Deed, 20 November 1849, WMSHS; "School Days"; Fowler, *Octagon House.*

[12]Seventeen-year-old John Watkins and two other boys were taking the second part of algebra, while sixteen of their schoolmates, including Kate (age 14) and Martha Watkins (11) and their cousin Sue Carter (21), were in the first part. Four students (Martha Watkins, age 11; Josephine Hollingsworth, 16; Samuel Wharton, 13; and William Wharton, 11) were enrolled in the same seven subjects. Franklin Academy 1859 Enrollments, WMSHS.

[13]*History of Clay and Platte Counties,* 271–79.

[14]*Liberty Tribune,* 30 March 1857.

[15]Perhaps following Watkins's lead, Professor L. W. Lewis, a 26-year-old native of Maryland, established the Lewis Institute in nearby Greenville in 1858. Along with neighbor J. C. Henneburgh, who specialized in teaching music, Lewis sought to elevate the cul-

tural level of the region. After the disruptions of the Civil War, he reopened the facility as Greenville High School. His offerings centered on literature and the arts for the youth of Washington Township. Of his location, the mentor averred, "We have selected this place away from the allurements to vice and immorality presented by the larger towns of our state among the green hills and in the grassy vales surrounding Greenville, nothing is presented save what tends to elevate and refine the feelings of Young Ladies and Gentlemen." Withers, "Education in Clay County."

[16]Franklin Debate Society minute book, 6 December 1842-3 January 1846, WMSHS; Potts, "Franklin Debate Society," 1–21; Wiebe, *Opening of American Society,* 166.

[17]"Old Men of Clay County."

[18]Franklin Debate Society minute book, 13 January 1843 and 8 April 1843, WMSHS.

[19]Historian Lawrence Frederick Kohl has emphasized, "The central problem of the Jacksonian era was to work out how these individualistic Americans would relate to each other." Franklin Debate Society minute book, 6 January 1843, 10 March 1843, 18 April 1843, and 29 December 1843, WMSHS; Kohl, *Politics of Individualism,* 12.

[20]Doyle, *Social Order,* 178, 12–13, 156, 99; Bender, *Community and Social Change,* 5–7.

[21]Clay County Circuit Court Records, Box 5, no. 385; Box 7, no. 537; Box 10, nos. 963, 1048; Box 12, no. 1338; Box 24, no. 3290; and Box 32, no. 3963, Clay County Archives; Clay County Commission Records, Book 7, p. 53, Clay County Archives; Petty, "Some Punkins," installments 19 and 20; "Old Men of Clay County."

[22]Clay County Commission Records, 12 September 1836, Box 3, p. 45, Clay County Archives; Clay County Circuit Court Records, 27 August 1836, Box 5, no. 332, Clay County Archives; *History of Clay and Platte Counties,* 123; *Illustrated Historical Atlas,* 10.

[23]Consistent with his contemporaries' tendency to stress personal character as the key explanation of success or failure, Watkins (in an autobiographical sketch done in his later years) chose to emphasize that "industry, integrity, and their practice made him respected and honored by his neighbors." He took particular pride in controlling desires, proclaiming, "Mr. Watkins is a man of temperate habits, having used neither spirits, tobacco, nor coffee for thirty years. He claims his success in life is due to his industry, economy, and promptitude." Squire Watkins was bourgeois and proud of it! O'Brien, *Grassland,* 154; Hyman and Wiecek, *Equal Justice,* 338; Franklin Debate Society minute book, 31 March 1843, WMSHS; "Old Men of Clay County"; *U.S. Biographical Dictionary,* 741.

[24]The designation of the area as Bethany persisted on maps through the era of the Civil War. "Bethany Post Office," in *Spinning Yarns;* "United States Post Office: Appointments of Postmasters," National Archives Record Service, Record Group 28, Clay County, MO; C. A. Evans, *Confederate Military History.*

[25]Barney, *Passage of the Republic,* 158.

[26]Bloch, *Paintings of George Caleb Bingham,* 1–32; *Liberty Tribune,* 27 February 1852.

[27]Haynesville was also, like many such villages, a center for churchgoers. The area was home to a number of Disciples of Christ; in 1852 the Disciples' founder, Alexander Campbell, visited the budding settlement for a revival. Dains, "Alexander Campbell," 18–21.

[28]Roney, *From Entry Fee,* 7–30; *Illustrated Historical Atlas,* 11, 28, 36.

[29]*History of Clay and Platte Counties,* 139, 262–63.

[30]Population Schedule, Eighth U.S. Census, 1860.

[31]Agricultural Schedule, Eighth U.S. Census, 1860.

[32]George Watkins to Mary Ann Watkins, 3 November 1855, George Watkins to Parents, 13 June 1856, WMSHS.

[33]George Watkins to Kate Watkins, 28 December 1856, WMSHS.

[34]George Watkins to Waltus L. Watkins, 9 November 1856, George Watkins to Mary Ann Watkins, 7 December 1856, WMSHS.

[35]George Watkins to Parents, 30 September 1856, George Watkins to Waltus L. Watkins, 18 December 1856, WMSHS.

[36]George Watkins to Waltus L. Watkins and Mary Ann Watkins, 22 February 1857; George Watkins to Waltus L. Watkins, 15 and 20 March1857, WMSHS.

[37]George Watkins to Waltus L. Watkins, 15 March 1857, WMSHS.

[38]George's grandmother, whom he was visiting at the time, was even more anxious and believed the Watkins family ought all return to safety in Kentucky "till the war was over but I told her there was no danger." George Watkins to Dear Parents, 13 June 1856, WMSHS.

[39]"The market revolution, as manifested in a bourgeois ideology that redefined works as the self-motivated virtue of economically rational individuals striving to improve themselves in open competition, gradually exposed slavery as an unnatural, and unnecessary, system of unfree labor." Watkins (who had never been a big slaveholder) seems to have been slow to act on this realization, as he was still holding a number of slaves in 1863. Barney, *Passage of the Republic,* 167; *Portrait and Biographical Record,* 216.

5

This Unholy War

Residents of Clay County, regardless of prominence, found it impossible to avoid the political whirlwind that swirled around slavery in the 1850s. The Democratic Party split over the issue, both in Missouri and nationally. Many proslavery Democrats turned against their former leader, longtime U.S. senator Thomas Hart Benton, who, though himself a slaveowner, opposed the extension of slavery into new territories; this position led to loss of his Senate seat in 1851 and defeat in several subsequent elections. The state convention to which Waltus Watkins was a delegate (1852) selected moderate Democratic candidates for the governorship and congressional seats, yet the situation continued to deteriorate. Watkins, who like many Missourians was tragically trapped between his proslavery heritage and his Union sentiment, would ultimately choose Unionism as he struggled to protect his family's interests during the Civil War years.

Border Warfare

Tensions flamed vigorously along the Kansas–Missouri border as the 1850s advanced. Following passage of the Kansas–Nebraska Act in 1854, which provided that residents of those territories rather than national legislators would determine whether or not to allow slavery, rival organizations arose to try to control the outcome. The Massachusetts Emigrant Aid Society (later reorganized as the New England Aid Company) sponsored settlers who envisioned Kansas as a free state. Proslavery forces in Missouri responded by organizing first the Platte County Self-Defensive Association and then Blue Lodges, to combat abolitionists, forestall free-soil settlers, and forcefully influence elections in Kansas. Long before the war started in April 1861, the clashes between such organizations, spurred on by men like abolitionist John Brown and Missouri's proslavery senator, David Rice Atchison, led to fearful and bewildering violence waged by "border ruffians," "jayhawkers," and "bushwhackers." Missourians, most especially those in the west, would suffer what historian Michael Fellman has labeled "the most widespread, longest-lived, and most destructive guerrilla war in the Civil War."[1]

Such partisan strife was rampant in Clay County.[2] In 1854–55, the Kansas Aid Society began to meet at the Liberty Courthouse to address the "emergency"; their goal was to furnish "outfits, etc., to persons desirous of immigrating to Kansas Territory and

Civil War broadside calling men to arms. Reproduced from a copy of the original (including the misspellings) WMA.

TO ARMS!

TO ARMS!

The citizens of Washington Township, and all others, conveniently situated, are earnestly requested to attend a meeting to be held in the town of Greenville, on Wedesday MAY 1, 1861, at 1 o'clock, P. M., at which time and place, there will be an effort made to organ-ize a Military Company, for the purpose of home defence, and the repelling of all hostile invasion. Let every on ebe present. Young, middle aged, and old, come! For the hour and the time demand that all take action. —

☛ Remember WEDNESDAY, MAY 1st.

also to determine the number of men to be sent with each wagon." They raised over $1,600 through subscriptions from thirty-six sponsors (including Waltus Watkins), who each pledged at least $50 "to serve the cause" of proslavery. In May 1856, armed with weapons and ammunition from the recently seized federal arsenal at Liberty, two hundred Clay County men assisted in attacking the free-state town of Lawrence, Kansas. When the war started, the extended Watkins family would be caught up in the turmoil.[3]

THE WAR BEGINS

Following South Carolina's secession on 20 December 1860, local leaders argued for moderation. Alexander Doniphan and O. P. Moss, both heroes of the Mexican War, sought to persuade their fellow Clay

Countians not to join the secessionist military ranks of Claiborne Jackson and Sterling Price. A rally was quickly organized in Liberty, at which reportedly several hundred people stood in the snow for three hours to hear Doniphan and other orators. Another local leader, *Liberty Tribune* editor Robert Miller, led his readers back from the brink of violence and seccession toward Unionism. In February–March 1861 a special state convention was held to determine whether Missouri would secede or remain in the Union. In spite of strong pro-Southern sentiments among many of its citizens, Clay County voted overwhelmingly to send Unionists to the convention, with only 4.9 percent voting for secessionists to represent them. Statewide the pro-Union faction prevailed, and Missouri did not secede.[4]

The Civil War came to Clay County only days after the Confederate attack on Fort Sumter, South Carolina, that signaled the official start of the conflict. On 20 April 1861, two hundred Southern supporters again seized the small federal arsenal at Liberty; "Minute Men" from four counties on both sides of the Missouri River stole away with firearms, guns, powder, and sabers in anticipation of taking law into their own hands. Within a fortnight, St. Louis was wracked by open warfare; soon thereafter Governor Claiborne Jackson and the state government created a pro-Confederate militia. A broadside circulating in Washington Township called for a 1 May rally at Greenville "to organize a Military Company, for the purpose of home defence, and the repelling of all hostile invasion." Forty-three men using the name "Washington Guards" turned out to drill under Captain L. M. Lewis.[5]

The same day, the *New York Tribune*'s Kansas City correspondent (a Unionist and abolitionist) reported, "Business is no longer here. Our most worthy citizens are leaving as rapidly as possible, and I am satisfied that within two months from this date our population will be reduced from seven thousand to two." The troubles locally could be traced to a "small but desperate and irresponsible band of Secessionists...trying to create a reign of terror" and seize control of the city. The local mayor, "a man of little ability, no principle, but a world of impudence," reportedly had led the traitorous attack on the Liberty arsenal. The onslaught of secessionists heralded the demise of "law and republican liberty" throughout the state.[6]

As the war ensued, most Missourians were, as Fellman puts it, "left confused and feeling caught in the middle of a battle they wanted to avoid. They remained loyal to the Union yet deeply resentful of Federal force." This resentment fueled the guerrilla actions that pervaded western Missouri throughout the war.[7]

Violence in Clay County

Although only one battle would be fought in Clay County (Blue Mills Landing, 17 September 1861), the war brought merciless ambushes, plunderings, and intimidation to homes throughout the area. Union occupation of Missouri left many Confederate sympathizers behind Federal lines. Pro-South irregulars, drawing on local sources of manpower and popular support, harassed Federal troops and terrorized Unionists, who did their share of marauding in return. Hearing of this, Mary Handy lamented to Waltus, "I am truly sorry for you all in Mo, truly so; & for your state also; who would have thought your noble and promising state would have been so sadly depridated [*sic*] in so short a time." Kate Watkins replied to her Kentucky kin, "You have never heard the half of the outrages that have been committed and you never will for it could not be told.... All the western border counties have been almost ruined by the Kansas jayhawks. The southerners have been the sufferers in almost every instance." Displaying her bias, she claimed that newspaper accounts of rebel atrocities were unfounded, but stated that "the jayhawks have stolen all the stock run off all the negroes and burnt a great many dwellings."[8]

Amid sustained guerrilla activity from both sides, the Union troops found it impossible to maintain control. Major General Henry Halleck, commander of the Union's Department of the West, lamented in 1862, "[The pro-Union jayhawkers] are no better than a band of robbers; they cross the line, rob, steal, plunder, and burn whatever they can get their hands on. They disgrace the name and uniform of American soldiers.... A few more such raids...will make [Missouri] as unanimous against us as is Eastern Virginia." On New Year's Day in 1863, President Lincoln wrote to the citizens of northern Missouri, conceding that regular Union troops had proved ineffective against bushwhackers: "Every county, every community, has been scourged, until scarcely a loyal family has remained untouched. Thus these desperadoes desolated the whole land, establishing a reign of terror." Privately he wrote his military commander in the state, "I am having a good deal of trouble with Missouri matters.... One class of friends believe in greater severity and another in greater leniency in regards to arrests, banishments, and assessments."[9]

The summer of 1863 proved tumultuous. At the beginning of the year, the President's Emancipation Proclamation had taken effect, though Missouri was not included because the proclamation only applied to territory behind Confederate lines. In July the Missouri State Convention voted to end the "peculiar institution," but not until 1870. Many slaves had already decided not to wait and instead took advantage of the turmoil (and, at least in the winter of 1861–62, of ice on the Missouri River) to leave their owners, adding to the chaos and bad feelings.[10] Tensions mounted with the burning of Lawrence, Kansas, by William Quantrill's band on the morning of 21 August 1863; the Watkins family feared that the "redlegs [Kansas troops] threaten to make Clay and Platte Co's. pay for it." As guerrilla activity increased and General Sterling Price threatened to invade Missouri, families in

Clay County faced the prospect of their sons being pressed into service by pro-South recruiters. Probably to avoid this possibility, John (age 22) and George (27) Watkins had, by February 1864, gone west to the Montana territory. (John was a teamster on the Oregon Trail; George settled in Montana and became a successful rancher.) Many of their Clay County colleagues went even farther, to Idaho.[11]

By 1864, the region's suffering neared a climax. As one observer lamented, "This once beautiful and peaceable land is forsaken and desolated, ruined, and only fit for bats, owls, & cockralls to inhabit." The summer of 1864 was more dangerous than any other time. The county history later recounted, "Now began that series of frightful scenes that occurred in Clay County in the summer and fall of 1864 when murders and killings were numerous, and robberies, plunderings and thefts were of such frequent occurrences as not to be mentioned...values were unsettled, everything was disturbed." Eighteen citizens of Clay County had already been murdered by one side or the other. In June, bushwhackers murdered a paroled Union soldier and a local militiaman. In August, the notorious pro-Confederacy raider Bill Anderson and his band stole arms, horses, and money from Union sympathizers. Liberty banker E. M. Samuel wrote to the state's military leader, "This country is in deplorable condition.... Is there no remedy for those who have through trials and sufferings, adhered to the flag of their fathers?" A newspaper editor wailed, "It is about time the authorities [in Washington]...should understand that the rebellion exists in Missouri, as actively, as virulently and as uncompromisingly as in Virginia.... In a word, the rebels hold the country, while the loyal people are besieged in the towns."[12]

Personal feuds often fueled the atrocities, and Waltus Watkins came perilously close to being a victim in one such episode. In June 1864, a Watkins Mill employee and his brother were murdered by Fletch Taylor, a bushwhacker leader who styled himself "Captain Commanding the County." Taylor sought to avenge the death of a man named Coffman at the hands of Union troops; he targeted Salmon G. Bigelow, probably because the Bigelow brothers "made themselves very conspicuous in their expression of Yankee sympathy." (Salmon had been a captain in a Federal regiment recruited in Missouri; John had mustered with a local militia company to fight desperadoes. Both were listed in the 1860 census as schoolteachers from Ohio.) One day, after Salmon returned home from Watkins Mill, he and John were confronted by a group of Southern sympathizers, who accused them of "some dastardly acts in the way of burning, stealing, murdering and running natives away from their homes." Both the Bigelow men were murdered. Jim Cummins, a horse thief and colleague of Jesse and Frank James (and possibly at one time a mule driver for Waltus Watkins), recalled the event forty years later, "I was an eye witness to the killing of these men, and, with some others, remember a little twelve

Watkins' eldest daughter, Kate, 17 years old in 1861, was an independent and self-confident young woman. She was also, for much of the war, an ardent Southern supporter, writing, for example, "I hope to see the time when the tyrants at Washington will not be troubled to take care of Mo. but she will live peaceably under the mild rule of Jeff Davis." Her letters to family and friends (signed "Secesh" to protect her identity if the mail were intercepted) described the impact of the war on Bethany and the rest of the county. She carried on a lively correspondence with her cousin Philip, who sent her accounts of camp life and of various battles in which the young soldier was engaged, as well as jesting remarks such as "The Ladies of the South thinks their is no body like the Mo. boys.... So the girls in Mo had better keep their eyes opened." Kate's letters also revealed her changing attitudes as the war dragged on and her zeal for the Confederate cause waned. It is not known to what extent her opinions reflected those of her father or the rest of her family, but the views she so vividly expressed were undoubtedly shared by many in western Missouri.[a]

Kate Watkins.
DNR, WMSHS.

[a]Kate Watkins to John G. and Mary Handy, 8 May 1861, WMSHS; Philip H. Gill to Kate Watkins, 18 January 1863, WMSHS.

year old daughter following up and bewailing the death of her papa and asking and praying what would become of us for the killing of her poor father." Federal Captain Rodgers and a militia company appeared at the crime scene soon after the bushwhackers departed, but Bigelow's widow claimed she did not know the murderers. "Captain Rodgers wanted to burn the factory and kill Watkins, but Mrs. Bigelow said no; that Mr. Watkins was a Southern man, but a good neighbor and friend." Apparently, even vengeance had its limits.[13]

Other events affected the Watkins family in far less dire ways. A schoolmate wrote to Kate from the Union bastion at Liberty that the county seat was so full of Federal troops "and doing so much mischief" she couldn't pay a visit. She went on to complain, "I say if I can't go where I want to, I can go to Kansas, like the darkies...I believe I had rather be studying that old Logic than stay here at home and see so many pretty soldiers."[14] Lizzie informed her sister that two "quite brazen young men" had robbed a stage she was on, perhaps on a trip to Lexington. Kate herself seemed to be taking things in stride as she replied to another correspondent, "You ask how the Clay Co. people are getting along, only tolerably well but I don't think times are quite as bad as the papers say. There have been some men killed, some houses burned, but we should expect no less." Less than a month later, she sounded less confident, confessing to her brother John, "Do you want to know something about political affairs and other news? Well I don't know what to tell you for there are so many reports that I can not tell the true from the false, a great many bushwhackers are reportd [*sic*] in Clay."[15]

Finally, in October 1864, warfare in western Missouri came to a turning point. On 23 October, Sterling Price's troops were repulsed by Union forces in the Battle of Westport; three days later "Bloody" Bill Anderson was killed. Within a week, Kate wrote to John that her sentiments had undergone a major change: "The rebels have been all through this country and have conscripted almost everyone fit for service....it is supposed that Price got at least 500 recruits in Clay County.... The bushwhackers have taken so much cloth that we were compelled to stop the factory, for they took it as fast as we made it and we thought they would stop when they got it all." The young woman who had called herself "Secesh" now used Unionist nouns—rebels, bushwhackers—to describe the raiders. Others apparently shared her disaffection; Kate reported that idle factory hands and local farmers in need of gristmill services were increasingly resentful of disruptions in their lives. She also blamed the Federals for two unexplained deaths of neighbors and attributed the illness of her uncle, Philip Gill, to "the present troubles." (Gill had been jailed for five days as a suspected Confederate sympathizer and had lost his son in the Confederate cause.) She went on to describe the martial law that had been in effect for six weeks: Men needed passes to travel to Liberty, for the Union forces occupying the town were wary of Confederate sympathizers. According to Kate, they were particularly suspicious of women; "they assign as a reason for this that a woman can learn more in a few minutes than a man can in a day."[16]

Perhaps the strongest indicator of Kate's change of heart was her statement that "Pres. Lincoln...is my choice of the two candidates" in the November 1864 election (in which, being female, she could not vote). In this, however, her neighbors apparently did not share her sentiments; only 206 voters in Clay County chose Lincoln, while opponent George McClellan polled 777 votes.[17]

The surrender of Confederate armies in 1865 did not bring about immediate stability in Clay County. In the November 1866 election, Clay County officials sought Federal forces from Fort Leavenworth to disband "an armed mob consisting of the most dangerous men in the county numbering more than a hundred." A similar group took control of the polls at Lexington in Lafayette County. The county would suffer from factional strife for a number of years to come.[18]

THE WAR AT BETHANY

The Family

The onset of the Civil War found the Watkins family scattered throughout the divided nation but pursuing many of their normal activities. The older daughters of Waltus and Mary Ann were in the border state of Kentucky visiting kin. They sought more money from home to prolong their visit (their lack of worry about the war perhaps indicated by their mention of a desire to tour nearby caves); Watkins, "scare of money," chided them for running out of funds so soon but promised to resupply them. George and John had hitched up teams and freight wagons and trekked westward on the Oregon Trail, stopping at Fort Benton and Virginia City, Montana, to sell farm and mill products. The four youngest children were at home, having passed school examinations, and were "doing many things to help" their parents.[19]

On the farmstead, despite a very wet spring and a grasshopper onslaught, "crops [were] looking tolerable well." In midsummer Watkins reported to his daughters that corn, wheat, and oats ripened in the fields, "a very good vegetable garden" featured abundant asparagus, and the cherries had been harvested in July. Despite constricting credit and a 25 percent drop in demand from 1860, Watkins had added carding machines within the factory. Proudly he announced, "Our new spinning jack has come at last and is nearly ready to start. It is a fine looking machine." In sum, his prospects appeared promising. Watkins withheld comment on the sectional struggle.[20]

Though a Southerner and prominent proslavery man before the war, Watkins seems to have remained fairly neutral once the fighting began. This prudent course may be the primary reason he, his family, and his mill came through relatively unscathed. Family records indicate that the Clay County provost marshall was satisfied with the Watkins family's loyalties to the Union. As a result, Watkins was able to secure the travel passes that he and his associates needed to conduct their affairs. A pass issued to John Watkins on 10 December 1862, permitted him to travel in five Missouri counties "for the purpose of buying sheep for his father who is enrolled as a loyal man."[21] Similarly George was able to haul farm produce to Fort Leavenworth. The family's loyalty status was enhanced by John entering his name on the rolls of the 43rd Enrolled Missouri Militia in 1862–63, though his widespread travels suggest the enrollment never actually led to military service. Waltus himself was considered sufficiently loyal to the Union to qualify as an official for the 1864 election.[22]

Correspondence for 1861-65 charted how the extended family—many members of which were considerably less neutral than Waltus's branch—fared in the war. Mary Handy wrote from Kentucky in 1862 that her son and a nephew had enlisted in the Confederate Army in "this unholy war," and that Meg Watkins Blythe, who lived in Platte County, had been victimized by "a band of outlaws & robbers." In the spring of 1865, Meg's husband, A. D. Blythe, was among those named as "Known Rebels and Bushwakers and Rebel Sympathizers" (identified as the Paw-Paw Militia) who were warned by the "Mid Knight Rangers" to depart or die. In November 1862, Jane Watkins Gill wrote that her husband Philip was harassed by two companies of pro-Unionists quartered in Haynesville, reporting, "Every time they have told me, that they had no charge against him, they only wanted to enrol his name, give bond, and take the oath; which was very foreign to his wishes or intension [*sic*]."[a] The Gills' son Philip (Kate's correspondent) was a member of the third regiment of the first brigade of Missouri Volunteers and saw action in at least three states; he was eventually killed in battle at Franklin, Tennessee. Susan Mary Carter confirmed the Gills's political leanings in her description of an 1863 visit: "We had a room full of the most ranting, low down secesh [women] you ever saw & such an up and cluck time as we did have." In the spring of 1863, the Gills were threatened by an axe-wielding slave, who was subsequently captured and punished with thirty-six lashings, leading Susan Carter to trumpet, "If he had been mine I would have sent him to prison & he might have stayed there until the moss growed on him ten feet long if Price did not get here before that long Hurrah for Price but d—I take the abolitionist."

[a]Mary Handy to Waltus Watkins, 16 March 1862; handwritten threat, c. 1864, WMSHS; Jane Gill to Dear Sister, 14 November 1862, WMSHS; Susan Mary Carter to Kate and Mattie Watkins, 22 March 1863, in Roney, "Old Watkins Mill," 146-49.

Watkins's political stance did not guarantee protection. A local history source related that "during the Civil War, Mr. Watkins was raided alternately by the boys in blue and the boys in gray." Local tradition claims Watkins hid his products in the woods to prevent theft by brigands or confiscation by Union patrols. Kate reported in sarcasm in June 1862 how the war had come home to Bethany: "The Feds condescended to pay us a visit although they were uninvited and unwelcome. They got in and were over half of the house before we knew they were on the place. They turned beds upside down searched drawers and trunks and joined and disputed around considerably.... [They plundered] 1 lb of powder from one of the boarders, took John's bullet mold and a stilletto from George." She was proud of Waltus's stance in the episode: "They told Father to come to headquarters [in Liberty] and prove himself loyal he could get the things back. F. told them he didn't want the things." The freedom with which Kate expressed her pro-Confederate opinions may be evidence that Waltus's loyalty to the Union cause was more pragmatic than heartfelt.[23]

The Watkins children's education, so valued by the family, was necessarily disrupted by the war. Kate's schooling was suspended for over a year because the schoolmaster, Robin Osborn, though a New Yorker, had joined some neighbors in the Confederate army. In the spring of 1863, Waltus and Mary Ann decided to send Kate and Mattie to Clay Seminary, a boarding school in Liberty, having apparently rejected another school in Platte City, closer to the Kansas border, out of fear for their safety. As Clay Seminary was run by a militant abolitionist, James Love, the move may be indicative of the Watkinses' neutrality. In their letters home to their parents, the girls hardly hint at the disruptions of war or enflamed sentiments; rather they are apprehensive about doing well in their studies and presenting a good image. Kate, right before June examinations, revealed that she was writing "under the scrutinizing gaze of Prof. Love, not that I will write any thing that I would not want him to see, but I always feel constrained so much that I cant express myself." She particularly wanted to be well dressed; among numerous items of clothing sought from her mother was "my organdy lawn dress," for her examinations. The school included two departments, Literary and Ornamental, and awarded the degree "Mistress of Arts." For reasons unknown, the Watkins girls did not receive their diplomas in the spring of 1864.[24]

In 1864, Watkins wrote his youngest daughters not "to be bothering your brains with the ups and downs of these Most troublesome times." Calling on the family religiosity to see them through, he reminded the girls, "We have all been placed in the Earth by a wise and beneficial Creator for some noble and wise purpose." Focusing on their purpose—represented by shared goals and values—helped the Watkinses bear the disruptions and deprivations of the war years.[25]

The Slaves

Loyalty to the Union did not prevent Watkins from continuing to hold slaves. In spite of a dubious claim in a later biography that Watkins "was never in favor of slavery, and set all his negroes free before the edict of the Government did so," it appears that his slaves were not freed, but either ran away or were emancipated by governmental decree. On the night of 13 August 1863, seven slaves then held by Watkins—five rented slaves, along with two that he owned, Lucinda and Ben—vanished, taking with them six horses and "a quantity of stolen clothes.... There was a general stampede the night they left, there is very few negroes left in the country." Only two slaves, the aged and enfeebled Parky and young Esseks, remained at Bethany, to be cared for by the Watkins family. (As a Missouri diarist noted, "Taking care of Negroes is a matter of humanity, not profit any more.") With the passage of the Missouri Emancipation Proclamation on 11 January 1865, these last Watkins slaves also received their freedom.[26]

Commercial Enterprises

By marketing farm produce to Union forces at Fort Leavenworth, providing custom carding to nearby farm families, and selling his factory products, Watkins prospered during the Civil War years. In early summer of 1864, Watkins announced, "I am more crowded with business this year than I have ever been." Eighteen wagons had been at the place at one time, some loaded with wheat, others with wool, "some of it to be carded, some to be spun & some to sell or swap for cloth." Watkins did not mention whether his patrons were Union or Confederate sympathizers, or both. An unsubstantiated local story says that he sold blankets from his broadlooms to both sides. Between mid-1863 and early 1865, the mill produced large lots of white flannel, which could be dyed in smaller lots per order, as well as gray flannel and broad linsey—all fabrics suitable for uniforms. There are, however, no army contracts among surviving factory records.[27] That Watkins apparently did not let politics interfere with his business relationships is further suggested by later reminiscences such as one published in the *Lawson Review:* "Many was the time that I heard...men speak of kindnesses received from Mr. Watkins, and of how honest he always was in all of his dealings with them."[28]

The war brought unexpected opportunities. William C. Sigon, an associate from Carrollton (forty miles away), relayed to Watkins in the summer of 1864 several reports that he was "dead, your factory and mill burned etc." Optimistically "proceeding on the idea that you still live and your factory is in successful operation," Sigon alerted Watkins that his market could expand, as other establishments—including the mill near Carrollton—were the victims of predatory guerrilla warfare or Unionist occupation. People in the region looked to Watkins "for their supply of winter clothing." He offered to amass raw wool from his neighbors' flocks to exchange for Watkins's fabrics.[29]

As the war ended in the spring of 1865, Watkins's newspaper ads showed a significant different in price quotations from previous years. In 1863, Watkins had charged 8¢ per pound for custom carding. Two years later, the work had risen to 12.5¢ per pound of plain wool or 15¢ for fine or mixed wool. He consciously sought to recruit customers from increasing distances as he emphasized his services would be rendered "with dispatch." Later, in 1868, he boasted he employed "a full corps of experienced workmen" and then had "on hand a full line of our celebrated make of Woolen Goods."[30]

Aftermath of the War

Race Relations

Following the Civil War, the Watkinses and others were faced with revolutionary social and demographic changes brought about by the freeing of the slaves. The racial composition of the population of Clay

County shifted dramatically. In the census of 1860, the Black population of Washington Township was 445 slaves and 9 "free colored," comprising 22 percent of the total population. A decade later, the census rolls recorded only 87 colored (3 percent). There is no record of Watkins hiring any of the remaining Blacks to work on the farm or in the mill. However, family correspondence for the next two decades is replete with domestic labor problems as free Blacks chose when, where, and how to work as servants. Carrie felt particularly put upon when Lucy, the Bethany cook, left her to go to school; she complained to Mattie, "I know she is the most ungrateful negro in the state. I feel like beating her good." Several years later Mattie, exasperated in her search for a competent cook, pronounced, "The servant question has gotten to be one of the greatest bothers of the age." Her husband, while traveling in 1891, remarked of racial adjustments in Virginia, "It is a hard lesson for them to learn to be without the negroes and the lazy shiftless things are in flocks, still justice compels me to say many of them are industriously improving themselves. But emerging from slavery, they show in their very progress toward improvement, how anxiously they copy what they consider evidence of culture in their former masters."[31]

Such remarks reflect common attitudes as the legal and social status of Blacks altered from slavery to freedom throughout the nineteenth century, a period in which three generations of the Watkins family interacted with Blacks in their households. Watkins's generation referred to "negro men and women" and frequently alluded to Blacks as members of a slaveholder's family. For example Mary Handy reported to her brother in 1843, "My family are all well at this time with the exception of my negro woman Lucy. She has a fine son 2 weeks old last Friday and is quite smart." (It should be noted, however, that at the same time the Handys regularly reported to their Missouri kin on prices for slaves, indicating Blacks were also viewed as a commodity.) Two generations and fifty years later, Waltus's granddaughter, during a visit to Kansas City relatives, profiled each white resident and concluded, "So there you have the family, except a lot of darkies, who don't seem to realize they aren't part of the family." Her description of one of her uncle's servants was unselfconsciously malicious: "I wish you could see Livy, our dusky eight year old Hebe. This is the way her countenance looks as she waits on the table. Her hair is tied up tightly with white twine and stands out like so many spears." Similarly, whereas in Waltus's era (1843) the Franklin Debate Society determined that Negroes rather than Indians had greater reason to complain of their treatment by the people of the United States, his son Joe reported of an 1875 a schoolhouse debate in the same neighborhood, "The question was weather [*sic*] a nigger had a soul or not they decided it that he had." The end of the century saw an increasingly segregated society where racial stereotyping seemed more profound.[32]

Frank and Jesse James

Lawlessness persisted in western Missouri in the postwar years. A Kentucky cousin wrote, "I see from the papers and hear from private individuals recently from your midst that more horrors are living in your unfortunate state. No human eye can penetrate or human tongue foretell the future, but have we not good reason to apprehend another civil war?" These "horrors" were personified by outlaws such as the legendary James Gang, many of whose members were former Quantrill guerrillas. The band led by Jesse and Frank James reportedly began its criminal career by stealing $60,000 from the Clay County Savings Association on 13 February 1866.[33]

The Watkins and James families shared a number of connections. As noted earlier, it was Frank and Jesse's father, Robert, who baptized Waltus in 1849; the two had been associates in founding William Jewell College and other Baptist endeavors. The family homesteads were only three miles apart, and the children were similar in age. It is probable that the James family purchased blankets from Watkins Mill, and that John Watkins later employed some of the James Gang as mule breakers.[34]

One Watkins family letter describes a notorious episode, the bombing of the James farmhouse by Pinkerton agents on 25 January 1875. Frank and Jesse's mother was severely injured and a younger brother killed in this raid. The next morning Lizzie Watkins wrote to Mattie, "A shocking affair occurred in the neighborhood, last night at Dr. Samuel[']s [James's stepfather]. Someone threw a kerosene ball into his kitchen. About one o'clock Mrs. Samuels and her little son got up and went to the kitchen to put out the fire. They stepped out on the porch, when they were fired at. She was shot in the wrist, but the boy's wounds were so fatal I suppose he is dead. He is said to be dying this morning. It is useless to conjecture who are the perpetrators of so horrible a crime but I have no doubt the 'James brothers' will make a desperate attempt to find out." When Mattie's father-in-law, Dr. James Scruggs, hastened to the James farm to treat Zerelda Samuel's wounds, his horse was stolen, perhaps in Jesse's escape. Sensation over "the Clay Co. tragedy" reverberated for months through state newspapers and the General Assembly.[35]

Squire Watkins

It appears that Waltus Watkins adopted a "soft" rather than "hard" stance in the rebuilding of community values in the period of Reconstruction. He did not want Confederate sympathizers, the disaffected, or Democrats punished. He sought to restore order amidst turmoil in ways such as recruiting stepnephew John Luke to reopen the school and to run Sunday school. He had no taste for the use of force; he did not join the militant Radical Union Party, with its tests of loyalty and punitive restrictions on voting and other liberties. In fact, although he had sworn a loyalty oath in 1861, he refused to take the "iron clad" test oath of 1865. The first action permitted him to keep his factory operating; the second meant he temporarily lost voting rights. Watkins, in 1870, claimed, "I refused to answer any but legal questions though he [the registrar] said he had never heard of my doing any unlawful act against the government." For Squire Watkins, the first was the act of a pragmatic businessman; the second a matter of honor.[36]

The Watkins family survived the most traumatic era in Missouri, the 1855–70 period. They suffered few real depredations, and no personal vengeance was inflicted on individual family members or their holdings. Their values and creeds proved resilient enough to withstand the challenges of the conflict. Watkins had protected his reputation, investments, and community standing. Now, advantageously positioned to capitalize on the upswing in the woolen industry, he could simultaneously continue to serve the needs of local clients with his other milling operations. Though he continued to be influential, he did not reenter the political arena.

Before 1860, the family-centered, landholding, evangelical values of rural Missouri farmers were under assault by changing marketplace values. Fellman has emphasized how folkways in Missouri were further disrupted by the guerrilla warfare, stating, "The Civil War was an unprecedented, enormous, immediate and brutal invasion of their communities in transition." At Watkins Mill this transition—symbolized by the factory—was accelerated as the war spurred demand for cloth and blankets. That the community around the mill was not totally devastated could be traced to Watkins's flexibility in adjusting to moderate scale manufacturing and in altering his political stance, both during the war and after. Along with his sons, Waltus accepted Unionism just as he adapted to industrialism.[37]

Comparing the censuses of 1860 and 1870 shows the impact of Watkins Mill on the neighborhood in the postwar years. Of the forty households near the Watkins homestead recorded in 1860, every head (including Waltus) claimed his occupation as a farmer, except for one miller and a carpenter. Ten years later, though farming still dominated, the mix now included a fireman, a carder, a railroad hand, a miller, two spinners, a dyer, a finisher, five factory hands, two superintendants, a wool picker, a brickmaker, three weavers, a fuller, and a blacksmith. Watkins now called himself a farmer and manufacturer. In 1860, only one of the forty heads of household had not been born in Missouri or a southern state. In 1870, southern heritage was still the norm in the sample, but Watkins's employees now listed birthplaces in Indiana, New York, Rhode Island, Ohio, Massachusetts, England, Ireland, Baden (Germany), Sweden, and Scotland. Clearly, the "factory on the farm" had brought a measure of cultural diversity to Washington Township—another social change in the period following the Civil War.[38]

Notes for Chapter 5

[1]McCandless, *History of Missouri,* 270–77; Fellman, *Inside War,* xvi.

[2]Violence seemed endemic to Clay County. Local jargon from the 1830s onward had its own term, "Mormonizing," for shooting, burning, or hanging one's adversaries. In 1854 frontier violence had flared at Smithville on the western edge of the county; three town leaders were murdered and three desperadoes hanged in an episode of vigilantism. *History of Clay and Platte Counties,* 164–66.

[3]*Liberty Tribune,* 8 December 1854, 28 June 1855, 20 July 1855, and 9 February 1855; *History of Clay and Platte Counties,* 174; Launius, *Alexander William Doniphan,* 236–37.

[4]Doniphan refused to join the fight and later, in 1863, moved to St. Louis. Fellman, *Inside War,* 8–10; Launius, *Alexander William Doniphan,* 244–52; Rold, "Secessionist Strength in Missouri," 417.

[5]The Clay contingent in the attack on the arsenal was led by Henry L. Routt of Liberty. Neither Waltus nor any of his sons joined the Washington Guards, but several members of the extended family did. Ethel Massie Withers, "Education in Clay County," Research Files, WMA Collection, WMSHS; DAR, *Centennial Souvenir,* 141; broadside, c. 1 May 1861, WMA Collection, WMSHS; *History of Clay and Platte Counties,* 197–201.

[6]*New York Tribune,* 8 May 1861, WMA Collection, WMSHS.

[7]Fellman, *Inside War,* 11.

[8]Mary Handy to Waltus Watkins, 16 March 1862; Kate Watkins to John G. and Mary Handy, 9 June 1862, WMSHS. In Kate's view, a temporary lull in June 1862 was due to the absence of pro-North troublemakers: "The ring leaders are not in Kansas. Jim Lane has been in Washington since Congress met, Montgomery is dead, Cleaveland was killed a few weeks ago and Jennison...was in Washington trying to get old Lincoln to make him Brigadier General of the Kansas volunteers and if he gets that office and is not checked Mo. will be almost in ruins this fall." Kate's hero at the time was "a man by the name of Quantrell [*sic*] who is fighting the feds on his own hook neither side claims him." Though "droves" of Federals were in pursuit, William Quantrill's band remained on the loose. See Goodrich, *Black Flag.*

[9]Major General Henry Halleck, quoted in Goodrich, *Black Flag,* 26; Abraham Lincoln to Citizens of Northern Missouri, 1 January 1863; Abraham Lincoln to Major General Samuel L. Curtis, 5 January 1863, in *War of the Rebellion,* series 1, vol. 22, 807–8, 3–8; Parrish, *History of Missouri,* 51–58.

[10]In 1863, a bitter correspondent wrote to Waltus's sister and her family, "Pa's negroes told him they were free now & he must look out for other help.... it is very nice to have yankeys sit up & free peoples negros, when they never had one in their lives. The poor darkies are the sufferers they will go to destruction as fast as the wheels of time can take them. I heard they were coming back from Kans. taking possession of rebel farms on the border." Mary (?) to Dear Friend, 5 January 1862, WMSHS; Anna Slayback to M. Blythe, [1863?], WMSHS.

[11]When measured against findings in a study of Jackson County, it is clear that Waltus Watkins suffered nominal losses and his sons resisted prevailing pressures to react strongly to military rule by Union forces. *History of Clay and Platte Counties,* 255; Roney, "Old Watkins Mill," ii, 113–33; Lizzie Watkins to Kate Watkins, 5 March 1864, WMSHS; Bowen, "Guerrilla War," 30, 51.

[12]Fellman, *Inside War,* 77; *History of Clay and Platte Counties,* 247; E. M. Samuel to Major General W. S. Rosecrans, 4 July 1864, in *War of the Rebellion,* series 1, vol. 41, part 2, 44–45; T. Dwight Thatcher, quoted in Goodrich, *Black Flag,* 127.

[13]Fellman, *Inside War,* 136–38; Cummins, *Jim Cummins' Book,* 91–93; *Liberty Tribune,* 1 July 1864; Craven, Interview. Cummins and the Bigelows apparently were neighbors; the 1860 census listed their households as being adjacent. Elsewhere in his book (p. 88), Cummins stated that the Bigelows "and a man by the name of Minter, all three of them offensive militiamen" had been killed by the James bunch.

[14][Mailiss?] to Kate Watkins, 24 July 1864, WMSHS.

[15] After a rival in the mill business, Weekly Dale, suffered an apparent stroke when pressed into protecting Liberty from the rebels, Kate wrote to her brother, "Had you heard that Mr. Weekly Dale was dead? He was struck with paralysis and died the next day.... His sons are running their factory but intend stopping when they get their warps out." Kate Watkins to John Watkins, 3 October 1864, in Means, "Edited Watkins Family Letters," 158–60; Lizzie Watkins to Kate Watkins, 18 October [1864?], Kate Watkins to Mr. A. C. Smith, 16 August 1864, Kate Watkins to John Watkins, 25 September 1864, WMSHS.

[16] In spite of her change in position, after the war Kate pledged $14.50 to the Southern Aid Society for sewing materials. Kate Watkins to John Watkins, 3 October 1864, in Means, "Edited Watkins Family Letters," 158–60; Bagg, "Kate of Watkins Mill"; pledge, 2 August 1866, WMSHS.

[17] Kate Watkins to John Watkins, 3 October 1864, in Means, "Edited Watkins Family Letters," 158–60; *History of Clay and Platte Counties,* 247.

[18] Parrish, *History of Missouri,* 142; Fellman, *Inside War,* 241.

[19] Waltus Watkins to Dear Daughters, 29 July 1861, WMSHS.

[20] Waltus Watkins to Dear Daughters, 29 July 1861, WMSHS.

[21] Proving loyalty required taking an oath of allegiance, as well as filing a performance bond. According to press reports, by the end of 1862 in Liberty district, 612 men had filed such bonds totaling $840,000, with individual bonds ranging from $1,000 to $10,000. Parrish, *History of Missouri,* 67; pass issued to John Watkins, 10 December 1862, WMA Collection, WMSHS.

[22] Kate commented, "Father, Mr. Finley and Mr. Williams are appointed judges, they think they will have a very peaceable election. I hope so." Kate Watkins to John Watkins, 3 October 1864, in Means, "Edited Watkins Family Letters," 158–60.

[23] *Portrait and Biographical Record,* 216; Kate Watkins to John G. and Mary Handy, 9 June 1862, WMSHS.

[24] Kate Watkins to John G. and Mary Handy, 9 June 1862, Kate Watkins to Josephine Hollingsworth, 1 September 1863, WMSHS; Kate Watkins to Mary Ann Watkins, 16 May 1863, in Means, "Edited Watkins Family Letters," 150–52.

[25] Watkins confessed, "I am exceedingly anxious that all of my children may do well and particularly my daughters...because you are more dependant [sic] than the male part of creation." Waltus Watkins to Dear Children, 2 June 1864, WMSHS.

[26] *Portrait and Biographical Record,* 216; Kate Watkins to Josephine Hollingsworth, 1 September 1863, WMSHS; Goodrich, *Black Flag,* 56; Meyer, *Heritage of Missouri,* 386. There is no record that Watkins pursued the runaway slaves.

[27] Waltus Watkins to My Dear Children, 3 June 1864, WMSHS; weavebooks, 1863–65, WMSHS.

[28] R. T. Clark, quoted in Roney, "Old Watkins Mill," i, 14. In a 1981 interview, Morgan Craven repeated accumulated local Civil War lore, including the belief that Watkins "had sold to both sides of the Civil War." Craven, Interview.

[29] William C. Sigon to Waltus L. Watkins, 30 July 1864, WMSHS.

[30] *Liberty Tribune,* 1863–68 clippings, WMA Collection, WMSHS.

[31] Carrie Watkins to Mattie Watkins, 18 October 1874, Martha W. Scruggs to Waltus L. Watkins, 22 November 1881, M. D. Scruggs to Dear Wife, 23 August 1891, WMSHS.

[32] Mary Handy to James M. Watkins, 28 May 1843, Mary Scruggs to Buford Scruggs, 3 and 11 June 1899, WMSHS; Franklin Debate Society minute book, 10 March 1843, WMSHS; Joe Watkins to Mattie Watkins, 28 November 1875, WMSHS.

[33] Stiles, *Jesse James,* 280–85; Settle, *Jesse James,* Parrish, *History of Missouri,* 139.

[34] Cummins, *Jim Cummins' Book,* 102–5.

[35] Lizzie Watkins to Mattie Watkins Scruggs, 26 January 1875, WMSHS.

[36] Potts, "Recollections"; Parrish, *History of Missouri,* 269; "Old Men of Clay County."

[37] Fellman, *Inside War,* xvii.

[38] Population Schedules, Eighth and Ninth U.S. Censuses, 1860, 1870.

6

A Reputation Not Surpassed by Any in the West

After more than twenty years of other commercial enterprises, Waltus Watkins decided, in the mid-1850s, to return to the textile industry. There is no written record of how or why he came to his decision, but a variety of factors may have come into play. He may have wished to provide his sons with training similar to his own apprenticeship. (George was 24 in the summer of 1861, when the mill opened, and John was 19.) He may have anticipated blossoming markets in nearby urban centers or in a wider region reachable through expanding railroad networks. Perhaps he felt driven to diversify his domain, to blend manufacturing with agrarian pursuits. The factors of political unrest and possible war might also have been part of his calculations. The 1858 opening of a woolen mill three miles north of Liberty no doubt opened Watkins's eyes to the possibilities. The owners of that factory, Weekly Dale and James McCoun, announced that they would manufacture wool into "anything suitable for the county"; they later advertised that they had procured the "services of Mr. Crowther, an excellent carder and manufacturer of wool" to run their machines. Whatever his reasons, it can be said that Watkins recognized and seized an opportunity for a new enterprise, in which he and his sons would reap moderate success for almost three decades.[1]

In some ways, becoming a "residentiary manufacturer" was a natural move for Watkins. His own training in cotton manufacturing and his experience with woolen processing in Liberty could be revived. Certainly he possessed the hands-on aptitude of the mechanic and recognized the potential of steam power. From the 1840s onward at Bethany, his business pursuits had furnished goods and services to customers within one or two days' travel. In addition, by 1860 he had amassed sufficient capital and credit to move into industrialism.[2]

What is surprising about Watkins's decision to start his woolen factory is the scale of his endeavor. He would build and equip a complete factory whose physical size and

The actual opening date of the mill is not known. One of Waltus's sisters wrote to another in February 1861 that "Brother Walters' large factory is not yet in operation, but will be by Spring."[a] In May 1861, Waltus still spoke of the mill in the future tense, predicting in an advertisement that the mill "will be in full operation in ten days," yet he claimed in the industrial census for the previous year that $8,000 worth of wool had been processed by a labor force of six hands. Other records indicate that 10,000 pounds of wool were carded in 1859–60 for use by home spinners. (Interestingly, the 1860 population census lists no carders for Washington Township, suggesting that those doing the work did not yet consider carding to be their main profession.) There is partial proof that complete factory operations commenced in August 1861. Thus it appears that carding and fulling services were offered starting in 1859 or 1860 (probably using animal-powered machines) and that the mill was fully operational in mid- to late 1861.

[a] Rebecca Yates to Meg Blythe, February 1861, WMSHS; Industrial Schedule, Eighth U.S. Census, Washington Township, Clay County, MO, 1860.

capacity made it one of the largest then in the Midwest. Within the three-and-a-half-story brick structure, a large steam engine powered a variety of fully integrated machines that could perform all of the major steps in processing raw wool into finished cloth, and do it in a matter of days at a price affordable to rural midwesterners. In so doing, he would personify not only the industrialization of the nation but also an ongoing philosophical question regarding the interface of agriculture and industry in the new republic.[3]

The Woolen Industry

Watkins's woolen factory is a manifold symbol of the Industrial Revolution and its effects on American life at all levels. During the period 1820–70, work that had previously been done at home or in nearby plots came to be performed in special buildings separate from home and of much larger scale than domestic structures. Artisans accustomed to using their own tools and honing their personal techniques were transformed into machine tenders or mill operatives who were paid a set wage. A large-scale energy source—the steam engine—drove an array of specialized and coordinated machines whose pace and productivity were seemingly superhuman.[4]

In the United States, early attempts to mechanize mass production were aimed at bringing Americans into a new role as purchasers of consumer goods. Commercial enterprise transformed the standards of living for nineteenth-century Americans and created a demand for items such as clocks, firearms, and textiles. The "pursuit of happiness"—largely measured by a growing economy—required a strong technological base, yet the emerging American republic suffered from a shortage of labor. The solution was to replace hands with machines. In the mid-1700s, textile production began to change, so that instead of being made by hand in the cottage or household, textiles were produced in factories. The first factories used the "putting out" system, wherein carded wool or cotton was sent to spinners and weavers working at home, and the woven fabric was carried back to the mill for finishing. A study of the transition around 1800 notes, "Even in an era when cloth was entirely handmade, the organization and structure of the craft was far more complex and diverse than previously recognized, as was the acquisition and consumption of textiles." As the process became more mechanized, it also became more centralized, with spinners and weavers working in the factory itself until eventually they were replaced by powered spinning and weaving machinery. Because cotton and wool each possess unique qualities, there emerged two distinct branches of the manufacture of yarn and goods for markets. By the time Waltus Watkins had learned his trade and moved to Missouri in 1831, the cotton industry had become completely mechanized; the woolen industry lagged somewhat behind.[5]

The first woolen processes to be mechanized were carding and fulling. (Detailed descriptions of these and other woolen production processes, as well as of the Watkins Mill machinery, appear in chapter 7.) Carding and fulling establishments offered services that saved home weavers the tedious and difficult work required to complete these processes by hand. Soon enterprising mill owners began to combine the two processes, adding new machinery as soon as it became practical for production use. The first half of the nineteenth century saw the introduction of machines intended to handle nearly every step in textile production. Although the cotton industry was employing power looms and spinning machines in the 1820s and 1830s, respectively, a commercially useful power loom for wool was not finally developed until 1841.

The 1810 census did not gauge clothmaking as a major industry, but by 1860 textiles constituted the largest industry in the nation, employing an estimated 165,000 men and women. The growing woolen industry could be measured by the increasing numbers of establishments in the business. In the years 1859–69 (the decade in which Watkins established his factory) the number of woolen establishments in the United States increased from 1,260 to 2,891, with the most noticeable gains being in the nation's interior. In the same period, the number of sets of mechanized carding apparatus and the consumption of raw wool more than doubled, and the number of employees nearly so. By the time of the nation's centennial in 1876, "woolen mills...could be found in every state in the Union, [in which] machines powered by steam or water and operated by semiskilled workers performed every major process of manufacture." Such factories quickly replaced all phases of home textile production, although

Above, workbench with wood lathe, blacksmith's drill, and vise; first floor south. Photo by Jet Lowe, 1978 (69).

Left, steam engine and boiler, "Washington Foundry, St. Louis, Mo." on condenser; in shed at east end of mill. Photo by Jet Lowe, 1978 (70).

skilled master weavers continued to produce and sell high-quality handwoven cloth well into the 1870s.[6]

As mechanization took over, the cost of cloth dropped dramatically. People could afford more, and better, clothing; curtains and other household textiles became common. By the last half of the nineteenth century, ordinary people could afford the fashionable clothes, elaborate draperies, and upholstered furniture that previously had only been available to the wealthy. Watkins Mill and mills like it all over America were part of this cultural evolution.

The woolen industry was noted for a great diversity of styles and therefore did not lend itself to the kind of large-scale operations common to cotton manufacturing. Cotton was usually woven in very long bolts of a few basic types, which were divided up for dyeing or printing in the latest fashions. Thus large amounts of cotton cloth could be produced in huge factories. Woolen goods, on the other hand, came in a much wider variety and had the patterns, and usually the colors, woven in. Factory owners who made too much of a current fashion might find themselves left with a lot of cloth they could no longer sell.[7]

Woolen Industry in Missouri

In general, industrial growth in Missouri lagged behind the commercial expansion that arose with the westward movement. State historian Perry McCandless explains, "Just when a factory system arrived in Missouri would be impossible to say, but the evidence suggests that some factory production was under way by 1840 with an increasing transition from pre-factory production methods after that date." Increases in population (to almost nine million in the eight midwestern states, according to the 1860 census) encouraged industrial expansion. Industrial historian Norman Crockett has observed, "As the density of the midwestern market and as per capita incomes grew, settlers in the region consumed larger quantities of manufactured goods and also demanded products of higher quality." The irregularity and high cost of transporting goods from eastern manufacturers gave entrepreneurs like Watkins the opportunity to meet the local and regional demand.[8]

As settlers moved to the western frontier, they brought their sheep with them, and the woolen industry followed. By 1860, Missouri was the site of eleven woolen manufacturers. These were scattered across the state, mostly in the more populated counties, with as much as seventy miles between mills. Only two were located in cities (St. Charles and St. Joseph). These wool-processing establishments claimed a total investment of $103,750; the same year, census records listed eighty-six wool-carding machines producing revenues of $300,000 (making Missouri the second most productive state in that field). These commercial operations relied in part on the 937,445 head of sheep then being raised by Missouri farmers. Crockett comments, "Midwestern wool manufacturers appear to have been opportunity-oriented rather than industry-oriented," meaning that Watkins and his peers were not particularly committed to wool but entered the industry because it happened to appear profitable. Indeed, the Missouri wool industry continued to grow, sustained in part by cloth and blanket sales during the Civil War. By 1878, according to textile directories, there were 126 woolen mills operating in Missouri.[9]

Crockett has profiled the average midwestern woolen mill in the 1870s as follows: It possessed two sets of cards, five looms, and 253 spindles; consumed 22,820 pounds of raw wool; and produced $18,000 worth of goods. The typical mill's workforce consisted of six men, three women, and a child, who worked for nine or ten months of the year. Power came from either steam or water and was valued at $17,733.[10]

By such standards, Watkins Mill was well above average. It boasted three sets of cards, twenty-four looms, and 864 spindles handled by upwards of thirty workers, who annually processed between 40,000 and 60,000 pounds of wool. For a brief period in the early 1860s, Watkins would claim his mill was the largest west of Cincinnati.

Figure 6.1: Comparative factory systems

CLASSIC FEATURES	RHODE ISLAND SYSTEM	WALTHAM SYSTEM
Ownership	Families and partnerships	Joint stock copporations
Capitalization	Often minimal	Large
Power at site	Less than 1,000 HP; often less than 100 HP	more than 1,000 HP; often much higher
Power distribution	Arranged independently by each mill	Power canal system shared by several mills or companies, often leasing power from a promotional agency such as Locks & Canals
On-site management	Owner	Agent
Labor	Children and families preferred	Young women preferred
Housing	Family houses and tenements	Boarding houses
Product	Emphasis on diversity	Emphasis on mass production
Sales	Manufacturer sold direct	Commission sales agency
Community form	Village	City
Example	Crown & Eagle Mills	Lowell

THE WATKINS WOOL FACTORY

Watkins's woolen mill generally followed the model established by Samuel Slater, a British immigrant who founded the American textile industry in the late 1700s. In this model, a family owned each firm, minimal capital was invested, and the owner personally supervised production. Family units of workers were employed and provided with housing in the hamlet. A diversity of products, rather than mass production, was initially emphasized, and sales were made directly to customers. Work was accomplished in batches rather than a continuous stream. As the industry evolved, the larger-scale Waltham (Lowell) system predominated, yet Watkins would stay primarily within the Slater model.[11]

Among the factors Watkins had to weigh as he pondered creation of a woolen factory were capital, machinery, raw material, labor, and the consumer market. These would be ongoing concerns throughout the active life of the mill.

Quest for Capital

Watkins's largest challenge was amassing sufficient capital to start and sustain his manufacturing venture. In 1870, the average midwestern woolen mill required between $17,000 and $18,000 to erect and equip; it may be assumed that costs in the early 1860s were on a similar scale. Depending upon who asked him and when, Waltus valued his factory investment from $10,000 upwards; in 1890, John Watkins estimated the valuation had increased to $30,000.[12]

Waltus combined a number of resources to secure the capital he needed. By utilizing brick and lumber from Bethany, he considerably minimized capital outlays in the initial construction. To increase his liquidity he sold four parcels of land in 1857–59. He turned again to his kin, traveling to Kentucky in 1857 and 1859 to negotiate with John Handy for loans. In 1860, he pleaded with Handy for $2,000 more; Handy tried to raise the money from local resources but with little success, as national political anxieties hampered extension of credit. Mary Handy alerted her brother back in Missouri, "I told him [John Handy] as you were in such a press, that as soon as he could get $500, he had better start it along, as that might save a suit, as well as cost & trouble." Ultimately Handy secured a $1,000 bank loan for his brother-in-law.[13]

Waltus also borrowed money from other relatives. His sister Rebecca loaned him $535 in 1859 and $1,162 in 1862; in 1864, he repaid $1,336.50. A promissory note for $1,162 was signed between Waltus and his mother in 1862. Additional land sales in 1864 and 1867 secured $2,900, perhaps for reequipping the factory. Finally, just as he allocated woolen mill profits to modernize the gristmill, so too did he shift moneys from the gristmill and other enterprises to support the woolen mill when needed. With funds from these various sources, he was able to construct, equip, staff, and maintain his factory.[14]

The Factory Building

Perhaps as early as 1855, Watkins had determined to erect a sizeable industrial structure at Bethany; in 1857, he directed his kilns toward making more bricks to meet the increased need. The structure, still standing at the Watkins Mill State Historic Site, measures 46 feet wide by 87 feet long. There is no foundation; the walls, 2 feet thick at the base, sit on a limestone shelf. There is a gradual taper to the walls; the walls on each floor are thinner by one brick than the one below. This technique provided a shelf for floor beams and joists, made walls lighter to bear, and economized on building materials. Tapered bricks were also used at the windows to gain additional natural light.[15]

The overall mill design is more typical of the 1820s than the 1860s. Watkins most likely looked to masons and other construction workers, perhaps experienced at the various building sites at William Jewell College, to help him design and create the factory; he may also have drawn on memories of the Kentucky factories of his youth. Structural expert Laurence Gross has observed that the configuration of the mill was neither as safe nor as efficient as textile factories then being built in the East. The inner space of Watkins's factory was not designed with concern for the flow from one stage of woolen processing to the next; instead the critical point was to get all machinery on the line shafts of the power train. Furthermore, he did not make use of advances then known as slow-burn construction (for fire safety), indicating that from the beginning his facility was out of step with the evolving industry.[16]

By 1860 the main building was complete (the most onerous work, removing topsoil to expose the bedrock foundation, reportedly having been performed by contracted slave labor).[17] Watkins promptly advertised the factory as "superior to any other in the western country," noting that "the house is very large and is calculated to run 36 Looms" (an overly optimistic goal, as it turned out). Over the next thirty-eight years the physical plant would be constantly changing as both the space inside the structure and the landscape outside were altered. Line shafts, pulleys, and belts had to be relocated. Looms were added on the second floor and an additional set of cards and new spinning jacks on the third. Major improvements, including a new well, came in 1868. About the same time, the gristmill was moved into its first-floor station to take advantage of steam power. A few years later, perhaps in 1876, the wooden engine shed at the southeast corner of the building was expanded to include a larger dye facility.[18]

A number of support structures evolved around the factory. Three ponds were dug to serve the mill (and required periodic dredging). An ample woolshed was built just across the Lane from the factory's west doors. To the north was erected a tall scalehouse where wagonloads of wool or grain would be weighed before storage and processing. To the south of the factory stood the blacksmith's shop. A small brick bleach house, southeast of the mill, provided space for removing foreign matter from the woven cloth. Further in that direction lay the cabin of William Waers, the dyer and bookkeeper. On the small ridge bordering the Bethany Road emerged housing for some factory workers.

Mill Equipment

Watkins relied on both local and national resources to furnish his woolen factory. He contracted in 1860 with the J. T. Dowdall Company's Washington Foundry in St. Louis to supply the boiler, steam engine, and flywheel, "and also furnish every thing necessary to have said engine complete & said engine is made in an approved stile & good workmanlyke manner." There were minor discrepancies between the contract and what was actually delivered. The boiler had four, rather than five, 12-inch return flues; the flywheel, weighing 8,000 pounds, was 16 feet rather than 10 feet in diameter. However, as the boiler, engine, and flywheel were to be housed in a shed adjacent to the factory, these discrepancies were easily remedied. The proposal, with a projected cost of $2,000, evidently included the understanding that Dowdall would secure an engineer (initially one Thomas Jones) to install and operate the steam power source for Watkins. Watkins accepted the proposal, put $1,000 down via the Branch Farmers Bank in Liberty, and awaited delivery before paying the balance. Over the next two years, he would pay over $3,300 plus shipping for equipment ordered from Dowdall; he would also attempt to recoup $800 from the foundry for shoddy workmanship and damaged goods.[19]

In July 1860, the boiler, engine, and some accessories, at a total cost of $2,738, were shipped aboard the steamboat *Thomas E. Tutt* from St. Louis to Richfield (Missouri City) via the Missouri River. Made of 28- by 78-inch hammered sheets hand-riveted together, the boiler had actually been salvaged from a sunken riverboat, and was somewhat dented as a result. Watkins had to wait until December 1860, when the ground was frozen, to haul the massive boiler the thirteen miles to the factory site (using, according to oral tradition, six pair of oxen to pull the sledge). It was set in brick with a nine-piece cast-iron firefront and grate bars fashioned by the foundry. This particular design, known as a return flue type or Scotch boiler, provided 560 square feet of heating surface. The heat and smoke from the firebox moved under the drum of the boiler, then up and back through the return flues to the chimney at the front. The fire, fed by four-foot-long logs, heated both the outside and the inside flues of the boiler. Although it was a reliable and relatively efficient boiler, soot buildup at the back of the return flues necessitated daily cleaning with a long brush or scraper. A lever-arm–type pop-off safety valve was originally adjusted to run at 100 pounds of steam pressure; after years of operation, the weight on the valve was shifted to reduce the pressure generated by the aging boiler.[20]

The engine powered the large flywheel (at 70–75 revolutions per minute) and two pumps. At first a double-action pump on the engine drew pond water to the boiler; later a ram pump drew from the deep well located inside the engine shed. An exhaust pipe vented steam into the factory to provide the necessary humidity levels for production. The engine was originally rated at 60 horsepower; later, when the boiler pressure was lowered to 60 pounds per square inch, the horsepower fell to 40. The speed of the 12-inch-diameter piston through its 36-inch stroke was regulated by a belt-driven Gardner fly-ball governor. Without the governor, the engine would speed up and slow down as the work load changed, causing the looms and other machinery to change speed as well. This would result in damage to the textile products. The engineer and a fireman were responsible for the boiler and engine operations and maintenance. In spite of their efforts, things could go wrong, as when upon occasion (during operation by less skilled workers) the single cylinder stopped on dead center; if that happened, attendants had to use pry poles to free it. In 1904, the packing, or steam gasket material around the piston, blew out. The engine has not run since.[21]

The main drive pulley, located inside the factory and attached to the flywheel by an octagonal shaft, measured 10 feet high, weighed 3,460 pounds, and had a 20-inch belt face. Drive train components were shipped in early 1861 so that by August the engine had carding production, at least, powered up and on line.

In February 1862, engineer Thomas Jones requested that the Dowdall Company ship another $992.18 worth of power train equipment via the Hannibal–St. Joseph railroad. With the addition of looms and spinning machines, the power source, flywheel, and main drive pulley now powered machinery on all three floors of the factory. Pulleys between the floors were connected by wide belts of leather or rubber-coated canvas; these turned the line shafts which, in turn, ran the mill's machinery. The 20-inch-wide belt face of the main drive pulley created much friction, thereby reducing the chance of slippage and power loss. The metal rod line shafts hung from the ceilings of the second floor and ran nearly the length of the building. Secondary jack shafts allowed some flexibility in machinery placement. When machinery was added or deleted, or changes in speed were required, the pulley and belt systems were adjusted.[22]

The production machinery was purchased from eastern fabricators and was, according to an early advertisement, "of the very best quality, of superior polish and finish." Spinning and carding machines were ordered from the noted firm Alfred Jenks and Son in Pennsylvania; the loomworks of Jenks, Stafford (Connecticut), and Furbush and Gage (Pennsylvania) supplied the various power looms. The specifications for these and other apparatus are found in Watkins's correspondence and in catalogs furnished by the manufacturers. Machinery costs usually totaled 70 percent of a wool manufacturer's capital investment, leading some midwestern pioneers of the industry to cut costs by purchasing secondhand equipment or by building machines from manufacturers's blueprints. Watkins, for example, built his own warping mill and dye vat apparatus. By April of 1863, Watkins Mill was completely equipped and fully operational.[23]

Operations

An April 1863 advertisement in the *Liberty Tribune* announced, "[The mill] is now in full operation, and although it has been running but a short time, has earned for itself a reputation not surpassed by any in the West."[24] The notice touted custom work plus a full selection of yarn, cloths, and blankets, and cited a network of "authorized agents" that included M. Royle in Lexington (Missouri), W. L. Woolfork in Platte City, E. P. Armstrong in Liberty, and Wason and McFarland in Richmond. As few business records survive from the initial

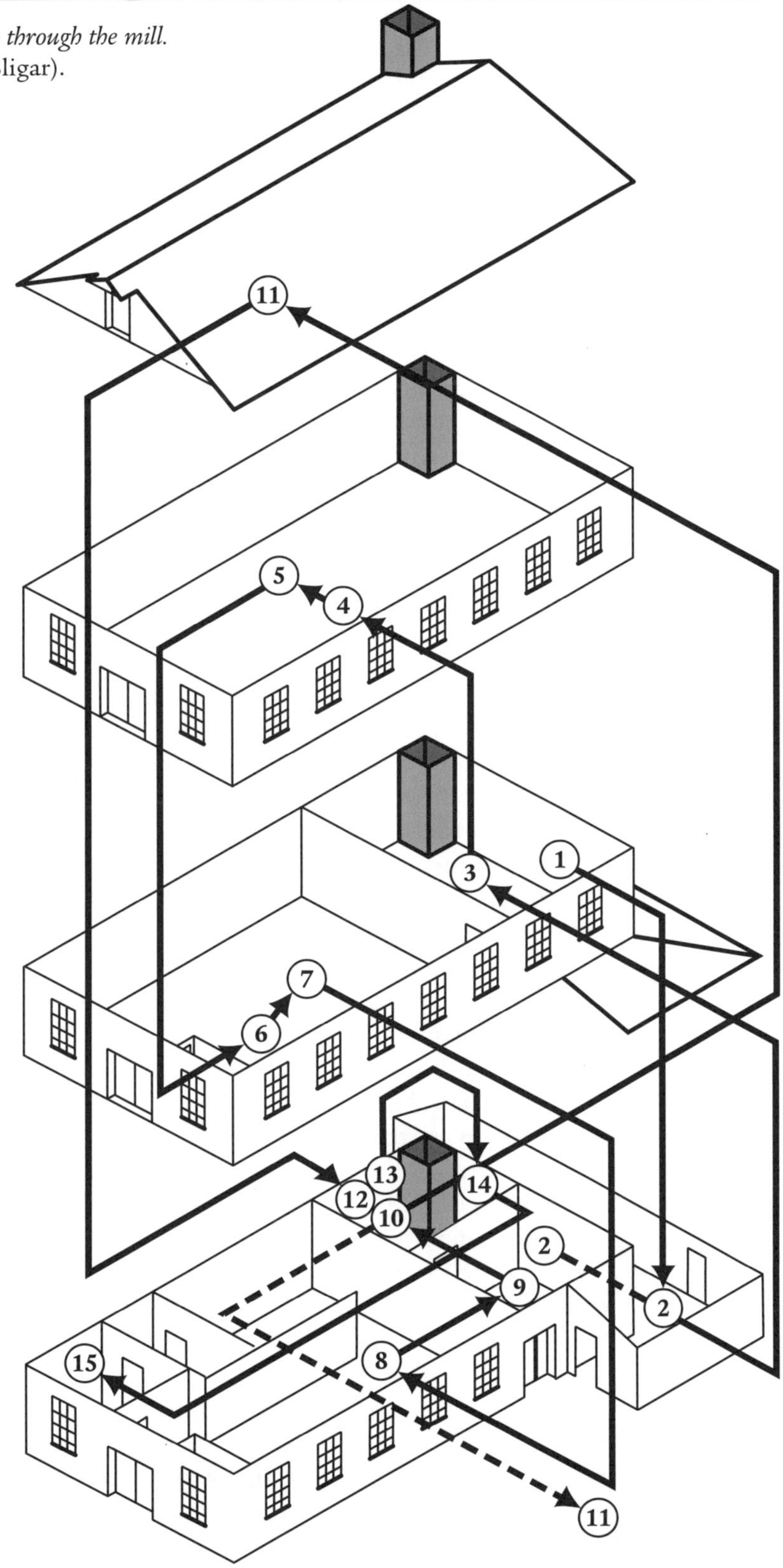

Cloth production flow through the mill.
DNR, WMSHS (A. Sligar).

1. Willowing
2. Scouring/Dyeing
3. Picking
4. Carding
5. Spinning
6. Warping
7. Weaving
8. Fulling
9. Washing
10. Gigging
11. Tentering
12. Brushing
13. Shearing
14. Pressing
15. Sales

years, it can only be estimated how often the machinery ran, what was the pace of manufacturing, and what were the outputs. Extant records indicate that between 1863 and 1886, Watkins Mill operated regularly, and that from 1886 to 1898, its services were used intermittently. (The gristmill remained in operation until 1904.) When it was in fairly continuous production, the factory operated eleven months each year, typically opening in mid-January and closing by 23 December. The average workday was eleven hours; there is some evidence that a handful of operatives worked as late as 9:00 P.M., though these were probably gristmill employees for the most part. The work week usually stretched from Monday through Saturday. At times Watkins, although a devout Baptist, might require Sunday work (with no such thing as overtime pay). Operating on the Sabbath opened Watkins to condemnation by church leaders, but he was never charged.

Watkins Mill operated under four successive sets of management and company names. From 1860 to approximately 1868, Waltus was the sole proprietor. In 1868, John, who had probably worked in the factory and had gone on at least one sales trip, joined his father as a partner in W. L. Watkins and Son. Then in 1882, John took in Judson and Joe to create Jno. Watkins & Bros., buying out Waltus, whose health was failing. An 1888 deed split the land holdings (then 3,670 acres) among the three brothers. After 1896, John, presumably the senior and dominant brother within the partnership, resumed sole control of the factory and half the land.[25]

Though the official name changed, it is likely that the establishment was always commonly known as "Watkins Mill." Long before the woolen factory existed, neighbors and customers would have grown accustomed to speaking of having their corn ground or their hickory sawn at "Watkins's mill." Waltus's bookkeeper, William Waers, always entered his records under the subject of "Watkins Mills" (plural).[26]

Extant business records for the firm include ledgers, production logs, weavers' books, employee payment records, correspondence, daybooks, and salesbooks. The discontinuous nature of many of these is partly due to the passage of time, but probably also reflects a lack of recordkeeping in the first place. There are many possible explanations for this omission: ignorance, trust, assumed knowledge, lack of attention to the details of the business. Custom work, because it required closer monitoring, produced more abundant records than did commercial production. Although incomplete records preclude definite statements about many of the Watkinses' business practices, it is still possible to draw some conclusions.

There is evidence of rudimentary efforts to estimate the costs of cloth production, thereby assisting in setting prices. Between August and December of 1872, Walthus kept track of production costs; as the next year began, he estimated cost per yard for three different cloths—flannel, cassimere, and jeans—from 100 pounds of wool processed in one week. Labor costs ranged from $37 to $44; supplies ran $9 to $12; contingence costs, covering insurance and machinery wear, were $5. Several errors in these calculations are now apparent; specifically, no cost was affixed to the raw wool, and the estimator anticipated that between thirteen and fifteen looms would be operating as opposed to the actual average of five to eight. Economist Paul Ogren has deduced that "the mill's commercial production, while not chronicled with the exacting precision of the custom work, did receive notice of a sort. Costs were estimated, weight or yardage loss calculated, and the products were precisely described." In the average year, Waltus's woolen mill produced $46,000 worth of goods, providing him about $4,000 in profit. In comparison, his profits from the gristmill averaged $3,000 per year. John, Joe, or Judson, all of whom had a smattering of college business courses, might have advised their father on these efforts of cost accounting.[27]

Raw Materials

Watkins had ready access to sources of raw wool. Throughout the Midwest, farmers found wool increasingly profitable and converted some of their resources from growing grain to raising sheep. Sheep did not cost much to sustain, were relatively cheap to feed, and did well in the comparatively mild winters in the lower Midwest. By 1870 midwestern farmers were raising thirteen million head of sheep, with Missouri as the fourth most productive state. In 1880, a booster of Missouri-grown wool reported in a trade journal that sheep raising in the state in the last decade had proved as profitable as any other agricultural pursuit. The article cited J. Vaughan of Lawson (a probable supplier to Watkins Mill) as a prominent example: "He said he had several thousand dollars loaned at 10 percent, and this business beat it about 200 percent."[28]

Some but not all of the mill's wool came from Watkins's personal flock. Before he started the mill, he had kept approximately 50 head of sheep, which provided meat for the family table and raw wool for Mary Ann and other women to spin. After his factory opened, Watkins increased his holdings to over 400 sheep. It is not known what breed of sheep these were. In light of Watkins's zest for highbred mules and cattle, it would not be surprising if some were merinos, a breed with exceptionally fine wool, yet the fact that much of the yarn and cloth produced at the factory was only of suitable or medium quality suggests otherwise. Perhaps Watkins's sheep were one of the mixed breeds of Spanish merino, French rambouillets, and English Southdowns with which Americans had been experimenting at least since the 1830s. Wool from such mixed breeds ideally suited demands for medium-quality cloth in the American market.[29]

Whatever the nature of his flock, Watkins could not provide adequate volume for his factory. He reported in the agricultural census of

1880 that he had processed 625 fleeces for 4,000 pounds of wool, yet at the height of its production, the factory consumed 60,000 pounds a year. To make up the deficit, Watkins turned first to local sheepherders. Perhaps he used the nearby services of O. F. Dougherty, the proprietor of Multonomah Farm just west of Liberty, who advertised himself as "Wool Grower and General Dealer in Stock and Trader of Same." When regional wool sources proved insufficient, Watkins bought additional wool from eight Missouri dealers and from interstate brokers such as Benjamin McLear in Kansas City and others as distant as Chicago and Philadelphia. He thus became part of a national market in raw wool.[30]

This market, as Norman Crockett's study has emphasized, was quite imperfect. First, the inherent variation in wool quality—two thousand distinguishable grades of raw wool, with as many as fourteen in a single fleece—made consistency difficult. Second, the system devised to meet wool demand contained many inefficiencies, yet competition from more efficient eastern suppliers was inhibited by the lack of transportation, especially the dearth of roads and railroads. Lastly, buyers and sellers had little means of gaining information on volume and prices in the annual shearing. Crockett infers, "The marketing of wool in the Middle West, therefore, rapidly became a speculative operation composed of many small buyers and sellers.... A feeling of mutual distrust between buyer and seller permeated the Midwestern wool market."[31]

As such market imperfections more often than not favored wool buyers rather than sellers, Judson Watkins met resistance from local agents during his circuits through the region as purchaser for the family firms. He also frequently faced competition from other mills or from a Kearney cooperative. He once thought a competitor, W. H. Arnold, inflated the wool market with the ulterior motive of currying favor among the electorate in his bid for public office: "He buys a little in a place at a high price and then goes on to another neighborhood." During his buying trips, Judson kept in touch by mail with John, who supervised production within the factory. In 1881, John instructed his brother, "Buy all the wool you can at low prices.... Some fool country buyers are trying to spring the market. We always had a stock of fools or smart allecks to contend with."[32]

Other raw materials for the mill included chemicals and dyestuffs, which were purchased from St. Louis and Chicago suppliers and originated in Central and and South America, Asia, and Africa. The cotton bagging for woolsacks came from nearby Kansas City.

Labor

As a skilled mechanic and experienced mill supervisor, Waltus Watkins knew well the need for trained labor in key positions within the factory. His recruitment of, and relationships with, dyers, finishers, fullers, and other craftsmen, as well as their less skilled relatives, are discussed in chapter 8.

Products and Services

In its heyday (1871-86) Watkins Mill produced a dozen different kinds of fabric, including flannel, jeans, cassimere, broadcloth, kersey, linsey, and satinet. Nineteenth-century jeans—not to be confused with modern cotton denim—was a durable twilled wool-and-cotton fabric ideal for rugged work clothes. Kersey was a twilled fabric for overcoats, trousers, and stockings. Cassimere (not the same as cashmere) was lighter, softer, dressier, and more appropriate for suits, as were flannel and broadcloth. Linsey and satinet were coarser, less expensive fabrics that combined wool weft yarns with cotton warps, as did jeans. Although worsteds—smoother, less scratchy types of woolen fabric—became increasingly popular in the latter half of the century, Watkins Mill did not produce them because of the specialized equipment and methods needed for their manufacture. Ogren, after reviewing voluminous sales receipts for the Watkins firms, has deduced that "the flannels, cassimeres and jeans, along with the yarns, formed the backbone of the product line. Even in the lean years all these items appeared. The cloths sold about equally well, with cassimere and flannel perhaps slightly stronger than the jeans." The mill aimed at a medium quality in the cloth market, leaving the finer goods to be purchased from eastern mills.[33]

To accommodate market tastes, various options were available. For instance, kersey was produced in sixteen different weights. Some fabrics were offered in patterns that included stripes, checks, herringbones, plaids, and small geometric figures. White, grays, browns, blues, and reds predominated in Watkins's cloth production, but various shades of green, lavender, and purple were also available.

The mill products were made to order. Stores, jobbers (wholesale dealers), and individual customers would specify the amounts, colors, and weights of the various fabrics they wanted, based on previous orders or perhaps inspired by samples provided by the mill. Once enough orders were received, Watkins would produce the required yardage of the requested products. Like other owners of small mills, he could not afford to turn out too much of any one kind. Working to order did leave him with periods of inactivity between jobs, but he was not saddled with an inventory of goods no one wanted.

The Watkins Mill product line also included yarn, shawls, and blankets. Yarn was offered in eight different colors (gray predominated), as well as a choice of weights. Blankets sold by the pound and came in three basic colors: gray, white, or white with colored stripes, although one extant example has a stripe composed of large oval daisies. Showing a flair for marketing, Watkins's advertisements boasted,

"My blankets will last as long as your lives and you needn't hurry about dying either."[34]

The custom services Watkins provided included carding, spinning, weaving, fulling, and dyeing. Demand for these services, which were primarily used by home weavers and others in the immediate region, fluctuated. In particular, finishing services such as dyeing and fulling showed much activity between 1863 and 1868 but fell off thereafter, as machine-made cloth replaced homemade. In the mid-1860s, the factory was handling between 1,600 and 2,200 custom jobs (primarily carding) a year. Then the demand decreased sharply until in 1875 only 513 such orders were processed. A brief revival occurred starting in 1884 as two mercantile firms, Hamacher Brothers in Richmond and Gallatin Dry Goods in Gallatin, forwarded to the mill orders that they had solicited. In Ogren's opinion, "The Hamacher orders possibly kept the Watkins family in business supplying nearly one-third of the total work taken in the years following 1886. The only other source of custom orders in these later years was the occasional walk-in customer in search of carding or spinning services." Eventually these last orders for custom work slowed until woolen operations finally ceased in 1898.[35]

Marketing

The mill's commercial production was aimed at selling a wide array of products on the textile market. Various sales schemes were attempted and a major change in product line effected in the lifetime of Watkins Mill. It was always possible to purchase yarns, blankets, or cloth directly at the mill. To reach wider markets after the Civil War, the factory initially (up to 1869) relied on a consignment system with merchants in the region. Watkins's salesmen solicited dealers who, operating on a 15-percent commission, ordered and sold a variety of cloth at prices established by the factory. Thereafter Watkins employed a few itinerant salesmen. Up to 1880, a sales staff that included Judson and John Watkins, Hiram Belt, and William Taylor traveled to Kansas City, St. Joseph, and St. Louis with sample books, order forms, and bolts of cloth for direct sales to individual customers, agents, and wholesalers. They called upon purveyors such as Royle and New in Lexington; Tootle and Hanna in Kansas City; and Bullene, Moore, and Emery in Kansas City. While collecting on bills or notes for such sales, they also purchased raw wool to haul back to the factory.[36]

The salesmen made as many as ten trips in a season (August to December). Hiram Belt recorded transactions in his daybook for a foray from 16 August to 14 September 1875: yarn sales netted $688.82 for twenty-five transactions, while twenty cloth sales garnered $345.91. Payments were made variously: $257.43 cash; $114.07 "to be paid next trip"; and $404.52 in sixty- to ninety-day notes. Belt calculated his charges at $40.08; his expenses at $27.80; and returned with $519.16 worth of raw wool (some as payment-in-kind for goods, some for cash,

The listings for W. L. Watkins and Son in the 1877 *Illustrated Historical Atlas of Clay County* categorized the firm's endeavors as "Manuf. of Woolen Goods, Miller, etc." It was a fitting description, and a noteworthy one, for up to that point Waltus had given his occupation, at least in census returns, as "Farmer." In addition to woolen manufacturing and custom work, the advertisement listed services as a flour and grist mill. (Interestingly, it did not specifically mention steam-powered milling operations, as did advertisers from Holt and Prathersville.) The Watkinses are also listed as farmers and stock dealers, with offices and salesroom located "near Lawson, Ray Co."[a]

A number of other commercial establishments, mostly those around the square at Liberty, also had listings in the atlas. No one in the area matched the Watkins family in the diversity of their enterprises.

[a] *Illustrated Historical Atlas,* 11–12, 14. The drawings also appeared in a similar atlas for Ray County.

and some for a $22.80 note he gave). On a subsequent trip, he left the factory with $569.21 worth of goods but was able to sell only half of his inventory.[37] By 1878–79, order books replaced the trip diaries submitted by salesmen. The more modern records charted pieces ordered at specified prices, as well as dates when the order had been filled. Usually orders were shipped within two weeks. By 1881, whole transactions were completed by mail, replacing traveling salesmen and signaling the advent of new, less personal sales techniques.

From 1876 onward, as the nation struggled with depressed times, cloth sales declined while yarn sales rose. John and Judson increasingly relied on the mails for finalizing sales of a more restricted product line that was marketed through jobbers located as far away as Omaha. A handful of accounts dominated. The factory increased yarn production in spasms during 1876–86 and 1889–92, but that market also soon dwindled. From 1890 to 1906, only on-site sales and a few specialty orders for either custom work or available supplies persisted.[38]

Competition

Although at the outset Watkins could anticipate a steady market in the region for his woolen services and products, he did not have a monopoly. Prior to 1860, carding and fulling services had been available to customers at Haynesville and Claysville. As noted, the McCoun-Dale operation opened in 1858; in about 1860, the owners moved the factory to Liberty, where it became known as the Dale Woolen Mill. During the 1860s, possibly due to war-driven demand for cloth, a number of other woolen mills appeared in Liberty and ran advertisements in the *Liberty Tribune.* By the time of the 1870 census, Watkins had at least five competitors in Liberty, among them one that combined a flour mill with its woolen factory. The proprietors of this establishment, the Corbin family, were mainstays of the town's banking and

WOOL CARDING!

The undersigned would respectfully inform the citizens of Clay and adjoining counties, that he is now prepared to CARD THEIR WOOL at the shortest notice. His Machinery is fitted up in the best style, with an entire set of NEW CARDS, and will be under the management of a gentleman who is a very superior workman (an Englishman) who learned the business where they learn it right. The particular attention of those who have fine Wool is requested, as I am prepared to do them ample justice (our cards being finer than usual.) All work done at this establishment

WARRANTED

to be done in a superior manner. The above cards will be run at my mill, five miles from Haynesville, and near the road from that place to Richmond.

PRICES FOR CARDING.

White Wool 8 cts per pound; Mixed wool 10 cts per pound.

Those who wish their wool manufactured into CLOTH, and who can wait until the first of October, can have their work done here, as preparations are being made to commence making the various qualities of Cloth in use by the Western people. ☞ The subscriber wishes to purchase a few thousand pounds of good Wool.

May 7th, 1860. W. L. WATKINS.

Right, broadside by Waltus Watkins, 7 May 1860.
DNR, WMSHS.

kinds of produce. People will find it to their interest to trade with him and build up the town.

MISCELLANEOUS.

B. F. Trimble, Plattsburg, Clinton Co., Mo., Breeder of thorough bred short horn cattle of approved families. Duke of Sharon at head of the herd. Stock for sale at reasonable prices. All are cordially invited to call and examine my stock. N. B. Duke of Sharon is a pure rose of Sharon bred by A. Renick who sells more short horns to go to England than any other breeder in the United States.—Address, B. F. Trimble, Plattsburg, Clinton Co., Mo.

S. C. Duncan, Proprietor of the "Bourbon Place," Smithville, Mo. Breeder of short horn cattle, Cotswold Sheep and Berkshire Swine. Best families and merits, choice young Heifers. For sale at all times and at lowest prices, also high grade of cattle of both sexes.—Address, S. C. Duncan, Smithville, Mo.

B. F. Record, M. D., Liberty, Mo., Physician and Surgeon, also Notary Public and D. D. G. M. of 13th Masonic District of Missouri.

B. T. Gordon, Liberty, Mo. Proprietor of the "Cottonwood Farm." Farmer, Stock Dealer and Breeder of short Horn Cattle and Berkshire Swine. *A Specialty made of Short Horns.*

O'F. Dougherty, Liberty, Mo., Pro. of the "Multnomah Farm," Farmer, Wool Grower and General Dealer in Stock and Trader of same.

W. L. Watkins & Son, Manufacturer of woolen Goods, Cassimeres, Flannels, Jeans, Blankets, Yarns, etc., etc., Carding and Spinning done to order, also Flouring and Grist Mill and Farmers and dealers in Stock. Office and salesroom near Lawson, Ray Co., Missouri.

M. M. Robinett, Merchant, Lawson, Mo. Dealer in Hardware, Furniture and Coffins, also a complete Stock of Lumber always on hand.

Samuel Schirtz, Claytonville, Mo. Physician and Surgeon. Calls promptly attended to at all hours.

Daniel Patton, Kearney, Mo., Minister and General Book Agent. Preached the first sermon ever preached in the city of Liberty.

A. P. Cutler, Holt, Mo., Pro. of Steam Flouring, Saw and Grist Mill. Manufactures all the best grades of flour and pays the best market prices for Grain, etc., also Farmer

Rev. L. J. A. Prather, Pratherville, Mo., Minister of the Gospel, Farmer, Stock Dealer and Miller.

George W. Woodruff, Pratherville, Mo. Proprietor of Steam Flouring and Grist Mill. Manufactures Flour and pays the highest market prices

Left, Watkins advertised in a list of Business References in the 1877 county Atlas. From the *Illustrated Historical Atlas of Clay County, Missouri.*

Below, Watkins was one of the "List of Patrons for the Atlas of Clay County" From the *Illustrated Historical Atlas of Clay County, Missouri.*

B. N. Moberley	"	30	Farmer and Stock Dealer	Madison Co., Ky.	1866
Geo. W. Woodruff	Pratherville	16	Miller	Pike Co, Ills.	1871

TOWN 53 NORTH. RANGE 30 WEST.

A. W. Gross	Lawson RayCo	10	Farmer and Stock Dealer	Clay Co., Mo.	1849
T. W. Gross	"	10	" "	"	1852
W. L. Watkins	"	10	Manuf. of Woolen Goods, Miller, etc.	Woodford Co., Ky.	1830
J. H. Watkins	"	10	" "	Clay Co., Mo.	1841
J. N. Craven	Prospect Hill	24	Farmer and Stock Dealer	Campbell Co. Tenn.	1856
B. Faucett	"	24	" "	Orange Co., N. C.	1843
Sallie A. Faucett	"	24	" "	Lafayette Co., Mo.	1872
Wm. M. Lynn	Claytonville	23	" "	Clay Co., Mo.	1841
M. J. Lynn	"	23	" "	"	1848
G. W. Madden	Prospect Hill	36	" "	Washington Co. Va.,	1846
J. F. Nichols	"	34	" "	Bourbon Co., Ky	1846
Wm. W. Crosset	Claytonville	28	" "	Orange Co., N. C.	1830

In his initial advertisement, published in the *Liberty Tribune* on 8 May 1861, Watkins promised a good assortment of woolen products at prices competitive with any west of Cincinnati. He noted that he still offered custom carding and other services but sought a more modern market: "I am of the opinion (and I can give a dozen good reasons for it) that I can give more satisfaction, and better bargains, by simply paying them [his custom carding patrons] for their Wool and selling them such Cloths as they may wish to purchase, than I could by working their wool for them." Any purchases at the factory could be made with cash (the more modern tender) or through traditional payment-in-kind of farm produce. With his eye on a particular type of consumer, Waltus stated his hope that "as many persons shall come and visit this as can conveniently do so, and particularly the Ladies, who are generally better judges than men, of such goods as I shall make."

merchandising. Among the competitors, Watkins was the only one who identified himself primarily as a textile manufacturer and listed his profession as such in the 1877 county atlas. It was a measure of his commitment to industrialism.[39]

In its prime in the 1870s and 1880s, Watkins Mill was the third largest woolen manufacturer in the state of Missouri, behind the Buell Manufacturing Company in St. Joseph and the Carthage Mill. Buell's entry in the textile directories listed eight sets of cards, seven spinning jacks, and forty-nine looms, compared to Watkins's three sets of cards, four jacks, and twenty-three looms; the Carthage Mill had 103 looms. Whereas Buell employed only male weavers and tended to produce higher-priced goods than those from Watkins Mill, Carthage largely employed women and children and produced cheap goods. In 1894, J. O. Buell offered to work as a superintendent under John Watkins, bringing with him plans to "encroach on their [Buell's] trade." (John was not interested.)[40]

Of the five woolen mills present in Liberty in 1870, only Jesse Bryant's factory was still in production in 1880. Corbin had returned to flour milling and the other three–James Adkins, Thomas Harper, and William Wrigley–had disappeared, reflecting the fate of the woolen industry throughout Missouri. Among their difficulties was a chronic cash flow problem stemming from the fact that wool had to be purchased at springtime clips, while most sales of yarn or goods were not made until late summer or fall. The Watkinses outlasted much of the competition, with the mill continuing to operate, albeit on a reduced scale, until 1898.[41]

"The Factory in the Field"

Waltus Watkins's woolen mill was a grand experiment, an attempt to reconcile industry and agriculture in a pastoral environment.[42] From Thomas Jefferson onward, America's leaders had feared technology as a political, social, and cultural threat to the young nation. Republican political structures were founded on agrarianism, with its principles of agricultural production by virtuous freeholding farmers. Technology, epitomized by machines and urban factories, stood in direct opposition to these values. Yet, as noted above, there was also a push to build the United States economy through production of consumer goods, which required technology. Politicians, philosophers, and ordinary citizens struggled with this ideological conflict.

For Jefferson, America's vast rural landscape offered a natural solution, as summarized by cultural analyst Leo Marx: "Once the machine is removed from the dark, grimy cities of Europe, he [Jefferson] assumes that it will blend harmoniously into the open countryside of his native land...helping to transform a wilderness into a society of the middle landscape." Others, including Alexander Hamilton, Tench Coxe, and Henry Clay, shared the view that "that by locating American manufacturers in the countryside and instituting a strict system of moral supervision, the health and virtue of operatives would be protected. Thus situated, manufacturers would harmoniously complement agricultural life, and the nation's agrarian character would remain undisturbed."[43]

For Waltus Watkins, there was no conflict. In the emerging "republic of technology," as Daniel Boorstin has labeled the United States, Watkins saw the factory and the field as complementary rather than contradictory. He found nothing incompatible about operating a major woolen factory in a rural setting and pursuing commercial farming and stock breeding on an equally large scale.[44]

Watkins's successful blending of agriculture, industry, and commerce is represented in two drawings that appeared in the 1877 *Illustrated Historical Atlas of Clay County.* Leo Marx has pointed to literature, painting, and architecture of the nineteenth century as grappling with the pastoral theme: "The pastoral form is a kind of deep structure. It sets up a yearning to resolve the conflict between an unmodified nature and civilization.... The pastoral is not to be equated exclusively with either technology alone or the natural world alone; the pastoral mode is an attempt to encompass both." As depicted in the atlas rendition, Watkins Mill had resolved this tension.[45]

The illustrations of the Watkins mill and farmstead occupied a full page; only two other county farmers and William Jewell College purchased such coverage. They were the work of an anonymous artist who, according to the publisher, had compiled and drawn his sketches "from Personal Examinations and Surveys" (probably in 1876) and who invoked his creative license in depicting Watkins's domain. The overall effect, reinforced by a balance between the residence and the woolen mill, emphasized the interdependency of the Watkinses' endeavors. The woolshed to the west of the factory and a line of workers' cottages to the northeast illustrated the industrial nature of the complex. (Other

Atlas illustrations of Watkins Mill (left) and Liberty Mills (below) show the vast difference in size and environmental setting seen by the artist. From the *Illustrated Historical Atlas of Clay County, Missouri.*

buildings, such as the blacksmith shop, the scalehouse, and the sawmill, were located beyond the picture's field of view.) Commerce was represented by two wagons on the Lane. Industrial power was symbolized by the steam and smoke drifting northward from the boiler's chimney and the engine's exhaust pipe. The artist stressed the general compatibility of animal husbandry and manufacturing by depicting horses, mules, swine, and cattle grazing in enclosures near the mill. Interestingly, except for the tentering rack shown to the right, no clues were provided as to the specific type of work going on within the factory building. In addition, there is no rendering of the quantitative or qualitative measures of Watkins's crops.

It is interesting that, whereas other farmsteads shown in the atlas are labeled with designations such as "Bourbon Place" or "Round Point Place," the name "Bethany" does not appear on either of the Watkins illustrations. Instead, the west end of the factory building features the more industrial-sounding "Watkins Mill." This is another example of artistic license, as subsequent investigations have uncovered no proof that the name was ever painted on the building as shown.[46]

The only other Clay County industrial plant depicted in the 1877 atlas was the Liberty Flouring Mills and Woolen Factory of O. H. Corbin and Company. Comparing the two illustrations underscores the nature of Waltus Watkins's operation as a "factory in the field." In the background of the Liberty Mill are symbols of technology (the railroad) and commerce (town structures)—a stark contrast to the bucolic and serene setting of the Watkins facility.[47]

The domestic side of the farm-factory blend is colorfully revealed in family letters to Mattie (Martha Watkins Scruggs), then a newlywed living in St. Louis, in the period 1874–75. The fall 1874 harvest was "better than we expected," especially the broomcorn. Waltus experimented in making peach butter; Lizzie claimed he "immediately took peach butter on the brain," and smiled as he stored apples in the cellar. Mary Ann made green tomato catsup and then turned to the seasonal hog killing. This year the scalding occurred at the mill, where the factory boiler provided hot water, rather than behind the house. Judson reported, "Mother has been trying to kill herself in the grease of eighty-five hogs so you know you will get all the sassage you can eat." Lucy, the cook, had left to go to school; still struggling with the cook problem over a year later, Carrie wrote, "I got so mad yesterday vowed I would marry the first fellow that asked me and try another field to see if it was so full of vexations as this."[48]

In mid-November 1874, a surprise 8-inch snowfall left 350 head of livestock to be fed. Joe had been tending to the chickens, cows, and hogs but it was John who lamented, "As usual on such occasions nearly all the hands want to get wood for themselves or do nothing." Waltus determined to put in a large crop in the spring, thereby making Joe "a regular field hand," and alternating Judson between factory and field. Spring chores included digging out the upper mill pond to a depth of fifteen feet, "cleaning up generally and moving machinery from one place to another better," and even scouring machinery in a boiling kettle.

Mary Ann tended her chickens and planted her vegetable garden in April 1875. After more than half the chickens died within a month, Lizzie warned, "You need not primp your mouth to get fried chicken oftener than once a week, this summer." Judson helped plant a hundred young apple trees at the home orchard, as well as at two tenant farms, and proudly reported that 300 acres of new ground had been broken up and corn planted in 160 acres of sod east of Lawson. Such optimism was quashed by a grasshopper attack in early June that devastated Bethany's gardens, orchards, and fields and led Waltus to write, "This is the darkest time that I have ever seen." So widespread throughout Missouri was the grasshopper plague that the governor was urged to create a bounty. He chose instead to proclaim a day of fasting and prayer. Watkins resourcefully responded by planting vegetables in boxes to be grown on top of the factory and plowing up an additional 280 acres in late June after the insects departed.[49]

"Manuf. of Woolen Goods, Miller, etc."

After two decades of building up his agricultural pursuits at Bethany, Waltus Watkins recognized and seized the opportunity to enter the woolen industry. The residentiary manufacturing unit he established—Watkins Mill—heralded change in rural Missouri as the habits, values, and mores of a traditional society were merged with those of an industrial culture. Its scale of operations, widespread use of machinery, and expanded markets provided a local example of changes affecting the economic system of the nation.

It may never be known if Watkins understood how unusual was his personal combination of endeavors (or whether he realized how risky it was to enter the woolen trade on the eve of the Civil War, itself a symbol of agrarian-industrial conflicts). There is, however, evidence that he was not wholly committed to manufacturing and that he did not view the woolen trade solely as a capitalistic venture. Rather, Watkins sought to speculate in the woolen industry, to experiment on a moderate scale in a manufacturing opportunity. Furthermore, he and John seemed to share the perspective that in the woolen factory, as in the sawmill and gristmill, the Watkinses were providing a civic service for a fee, a paternalistic viewpoint that occasionally conflicted with the profit motive. Although the woolen profits might reach a maximum of $10,000 per year, manufacturing never displaced agrarian pursuits as the family's primary endeavor. Such duality can be differently interpreted. Ogren has concluded that "there was co-existence, perhaps based on enterprise and diversification, but also based in conservatism and timidity, that made the Watkinses the manufacturers/agriculturalists they were." Alternatively it could be argued that the Watkinses were not timid but opportunely invested in a profitable sideline and had sense enough to minimize losses when that opportunity vanished. Whatever the truth, Waltus Watkins's approach to his entrepreneurial activities had profound implications for how he chose to operate his woolen factory.[50]

NOTES FOR CHAPTER 6

[1] *Liberty Tribune,* 20 May 1858 and 2 September 1859.

[2] Crockett, *Woolen Industry,* 23, 77, 129.

[3] *Homespun to Factory Made,* 66–72; Gross, "Wool Carding"; Cole, *American Wool Manufacture,* 1:351–67.

[4] Henretta et al., *America's History,* 293; *Homespun to Factory Made,* 54, 102.

[5] C. Clark and Hewitt, *Who Built America*, 371–75; Hindle and Lubar, *Engines of Change,* 185-204; Hood, *Weaver's Craft,* 1. Readers interested in the mechanization of textile production and the role textiles have played in American history are directed to the online exhibit "Whole Cloth (Society for the History of Technology).

[6] Increased textile production contributed to the overall industrial growth of the country in this period. The total American industrial output in 1810 was $162 million; the 1860 equivalent (in constant dollars) was $2 billion. Per capita output grew from $22.60 to $60.61. Hindle and Lubar, *Engines of Change,* 165-66; Cole, *American Wool Manufacture,* 1:84–87, 1:378, 2:39; *Homespun to Factory Made,* 102.

[7] The Bay State Corporation, built in Lowell, Massachusetts, in 1848, was the largest woolen mill in the U.S. and employed close to two thousand people. In spite of its size—or perhaps because of it—the mill went bankrupt within nine years. *Homespun to Factory Made,* 102.

[8] In 1860, slightly more than three thousand manufacturing establishments contributed to the Missouri economy. These firms claimed a capital investment of over $20 million, produced $24 million worth of products (compared to $8.5 million in 1850), and employed twenty thousand workers. The clothing industry was relatively small, contributing only $500,000. McCandless, *History of Missouri,* 152–53; Crockett, *Woolen Industry,* 28, 95, 40, 129–30.

[9] Ogren, *Watkins Mill,* 1–4; McCandless, *History of Missouri,* 153; Crockett, *Woolen Industry,* 30. In 1876, the number of woolen mills listed was forty. The size of the increase—more than 300 percent in two years—may be attributable to the fact that some firms did not bother to enroll in the directories, and many were vague on their production, making the directories somewhat unreliable as sources. Still, the directories provide an overview of the number of firms and their production, especially of the rapid decline after 1882. Matthews (Sligar), "Woolen Trade."

[10] Figures for Missouri's woolen establishments were slightly lower. Crockett, *Woolen Industry,* 79.

[11] The Slater (Rhode Island) system was one of four major patterns identifiable in American textile manufacturing as it developed on the East Coast in the late eighteenth and early nineteenth centuries. The patterns are defined primarily by variables of capital, energy sources, mechanization, and labor. A second model was the Waltham system, developed by a group of Boston investors, most prominently in the massive mills at Lowell, Massachusetts. Its chief characteristics were large amounts of pooled capital, corporate ownership, management by hired agents, an integrated facility with a large power source, and an emphasis on mass production. The Brandywine system, similar to the English, was employed in Pennsylvania and Delaware and featured a widespread exchange between small mills and individual workers—often at home—who produced specialized products. The fourth model, the Southern, comprised the southern cotton and woolen mills that arose after 1840, in which the relative productivity of free or slave labor was a key question. Dunwell, *Run of the Mill,* 52; Hindle and Lubar, *Engines of Change,* 185, 202–4; Maier et al., *Inventing America,* 309–14.

[12] Crockett, *Woolen Industry,* 32; Industrial Schedules, Eighth-Twelfth U.S. Censuses, 1860-1900; Roney, "Old Watkins Mill," i, 76–78. Evelyn Petty ("Some Punkins," installment 34) claims that by 1859 "the county [was] in a highly prosperous condition.... Money was plentiful."

[13] Mary Handy to Waltus Watkins, 24 September 1860 and 2 December 1860, WMSHS.

[14]Promissory note, Waltus Watkins to Jane Watkins, 11 April 1862, WMSHS; Roney, "Old Watkins Mill," i, 96–106; Ogren, *Watkins Mill,* 38.

[15]The number of bricks in the factory building is estimated at 450,000, which was four to five years' output from Waltus's kilns. Working backward from the mill's opening in 1860 gives a probable construction startup date in 1855 or 1856. "Mill Architecture," interpretive manual, WMSHS.

[16]Gross, "Importance of Research," 24–25; Gross, Audiotape recordings, tape 1.

[17]The reputed use of slave labor in the mill's construction has not been documented. Alterations to the factory building were made in about 1868 to accommodate the gristmill, which was moved inside around that time. The changes included removing the brick floor and erecting support beams. Interpretive manual, 1999, WMSHS.

[18]*Liberty Tribune,* 8 May 1861; Dagley, Interview; Rucker, "Archaeological Investigations."

[19]Waltus Watkins to J. T. Dowdall Co., WMSHS; Slates, "Watkins Mill Steam Engine." L. H. Jones and Jesse Walker were the engineers in the 1870s.

[20]On an average day the boiler's fire consumed two and a half to three cords of wood. Thus if an acre of the adjacent woodlands yielded thirty cords of firewood, then Watkins's steam engine deforested 20 acres per year. Shipping order, 25 July 1860, WMSHS; Matthews (Sligar), "Engineering Treasure."

[21]Dagley, Interview.

[22]Matthews (Sligar), "Engineering Treasure."

[23]*Liberty Tribune,* 8 May 1861; Crockett, "Westward Movement."

[24]*Liberty Tribune,* 24 April 1863.

[25]Ogren, *Watkins Mill,* 14–15.

[26]Watkins Mill Business Records, Jackson County Collections, WMSHS.

[27]Ogren, *Watkins Mill,* 96; Matthews (Sligar), "Engineering Treasure," 15.

[28]Agricultural Schedules, Seventh and Eighth U.S. Censuses, 1850, 1860; *Homespun to Factory Made,* 6–8, 56–59; Archer, "Wool Growing in Missouri."

[29]Agricultural Schedules, Seventh and Eighth U.S. Censuses, 1850, 1860; *Homespun to Factory Made,* 6–8, 56–59.

[30]Agricultural Schedule, Tenth U.S. Census, 1880; *Illustrated Historical Atlas,* 14; Ogren, *Watkins Mill,* 37–39.

[31]Crockett, *Woolen Industry,* 40, 42, 51.

[32]Judson Watkins to John Watkins, 29 May 1886, John Watkins to Judson Watkins, 10 June 1881, WMSHS.

[33]Ogren, *Watkins Mill,* 105.

[34]*Liberty Tribune,* 1861–65.

[35]Ogren, *Watkins Mill,* 86–87.

[36]Ogren, *Watkins Mill,* 99–103, 113–14.

[37]Hiram Belt daybook, 1875, WMSHS.

[38]Ogren, *Watkins Mill,* 114–15.

[39]*Illustrated Historical Atlas,* 11, 14, 33; *Liberty Tribune,* 1850–80; Population Schedules, Eighth, Ninth, and Tenth U.S. Censuses, 1860, 1870, 1880.

[40]J. O. Buell to J. H. Watkins, 8 November 1894, WMSHS; Matthews (Sligar), "Woolen Trade."

[41]*Illustrated Historical Atlas,* 11–12, 14, 33.

[42]After the political (republican) revolution of the 1760s–80s, Americans then faced another: the industrial revolution. In Watkins's generation, advocates would argue that "manufactures, as well as agriculture, harnessed natural resources and fulfilled nature's purpose." Like Jefferson, Watkins followed a decentralized plan for manufacturing. His factory and farm operations fulfilled such a blended vision in a rural, not urban, setting. Kasson, *Civilizing the Machine,* 19–28.

[43]Marx, *The Machine in the Garden,* quoted in M. R. Smith and G. Clancey, *Major Problems,* 122; Kasson, *Civilizing the Machine,* 62. Laurel Thatcher Ulrich *(Age of Homespun)* notes the complementarity of homespun and factory-made textiles in the nineteenth century.

[44]Boorstin, *Republic of Technology.* Adrienne Hood emphasizes regional variations in adjusting to industrializing textile production and distinctive urban and rural responses, and states that "manufacturing and agriculture were collaborative enterprises, each affecting the other and setting the terms for indigenous textile production." Hood, *Weaver's Craft,* 155.

[45]*Illustrated Historical Atlas,* 42; Marx, *The Machine in the Garden,* quoted in Lee, "Paradise Limited," 37.

[46]The picture also inaccurately represents the name of the firm, which in 1877 was W. L. Watkins and Son. Another point of difference between the Watkins illustrations and others is in the inclusion of sketches of the owners with their spreads; no such portraits of Waltus and Mary Ann appear in the atlas.

[47]*Illustrated Historical Atlas,* 33.

[48]Carrie in fact never married and remained at the homestead until 1943. Family legend claims Waltus disapproved of Carrie's initial suitor. Carrie Watkins to Mattie Watkins, 18 October 1874, 7 November 1875, and 11 May 1876, WMSHS.

[49]Various to Mattie Watkins, 1874–76; Waltus Watkins to Mattie Watkins (Mrs. M. D. Scruggs, 7 June 1875, WMSHS; Christensen and Kremer, *History of Missouri,* 9.

[50]Ogren, *Watkins Mill,* 11.

1. Above, display of Bethany farm tools at the Visitor Center. DNR, WMSHS.

2. Left, the history of the WMA is shown in this display case at the Watkins Mill Visitor Center. DNR, WMSHS.

3. Below, a crowd looks on at the 1958 auction of Watkins Mill. DNR, WMSHS.

4. *Left, the fruit drying shed and tra*
were used to preserve food for sale ar
use in the Bethany household. DNR.
WMSHS.

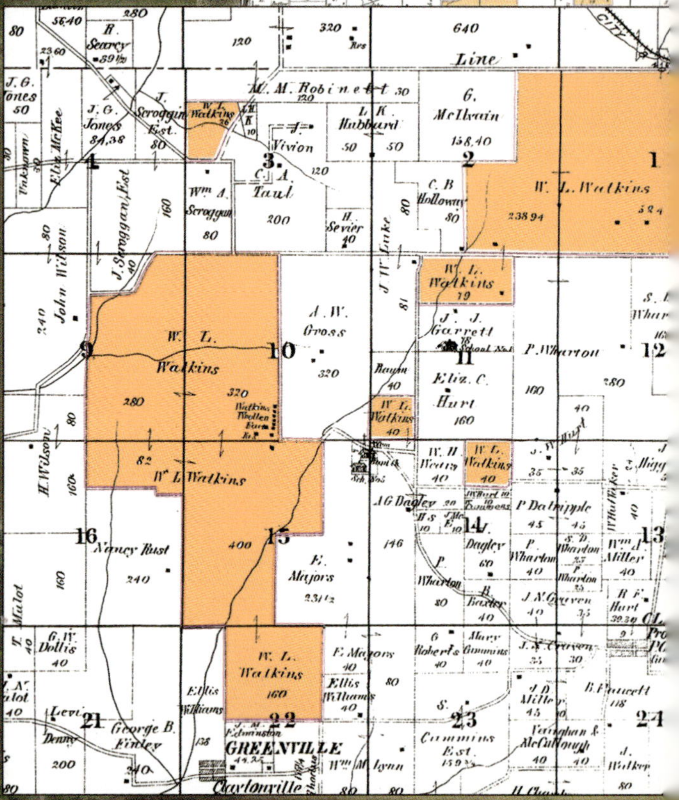

5. *Right, overlay of section map showing land owned by Waltus L. Watkins, 1877.* Illustrated Historical Atlas of Clay County, Missouri.

6. *Below, abundant gardens and orchards on the Bethany farmstead supplied the kitchens with fresh vegetables and a source of income.* DNR, WMSHS.

7. Right, sheep at Bethany were sheared in the spring and the wool used in the mill. DNR, WMSHS.

8. Below, Bethany farmstead was home to several varieties of turkey, geese, and chickens. DNR, WMSHS.

9. Bottom, merino sheep graze in one of many fenced pastures at Bethany. DNR, WMSHS.

10. Right, the north side of the mill in its run-down state, c. 1920. DNR, WMSHS.

11. Below, samples of woolens produced in the mill. DNR, WMSHS.

12. Below, diagram of woolen mill departments and processes, viewed from southeast corner of mill. DNR, WMSHS. Drawn by A. Sligar.

1. *Scouring and dyeing*
2. *Willowing and picking*
3. *Carding*
4. *Spinning*
5. *Yarn production*
6. *Warping*
7. *Weaving*
8. *Fulling*
9. *Washing*
10. *Finishing*
11. *Tentering*
12. *Pressing*
13. *Engine and boiler*
14. *Flywheel and main drive*

13. Right, a spinning jack, also called a hand-jack, produced uniformly spun yarn that could be woven into high quality cloth. DNR. WMSHS.

14. Above, the twister produced the heavier multiple-ply yarn used for knitting and some weaving. DNS, WHSMS.

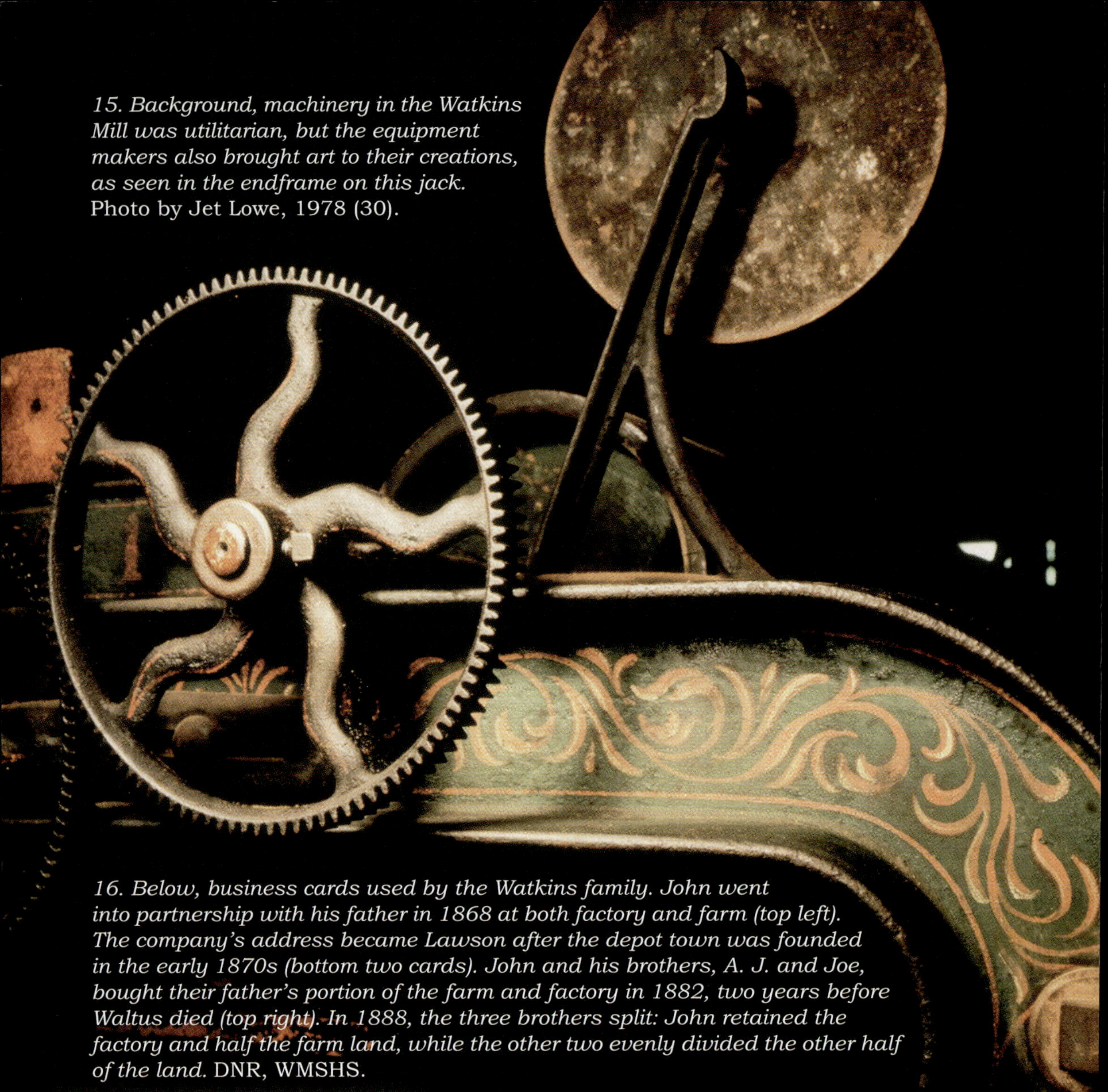

15. *Background, machinery in the Watkins Mill was utilitarian, but the equipment makers also brought art to their creations, as seen in the endframe on this jack.* Photo by Jet Lowe, 1978 (30).

16. *Below, business cards used by the Watkins family. John went into partnership with his father in 1868 at both factory and farm (top left). The company's address became Lawson after the depot town was founded in the early 1870s (bottom two cards). John and his brothers, A. J. and Joe, bought their father's portion of the farm and factory in 1882, two years before Waltus died (top right). In 1888, the three brothers split: John retained the factory and half the farm land, while the other two evenly divided the other half of the land.* DNR, WMSHS.

17. *Right, 1867 Bridesburg catalog used to order machinery for Watkins Mill.* DNR, WMSHS.

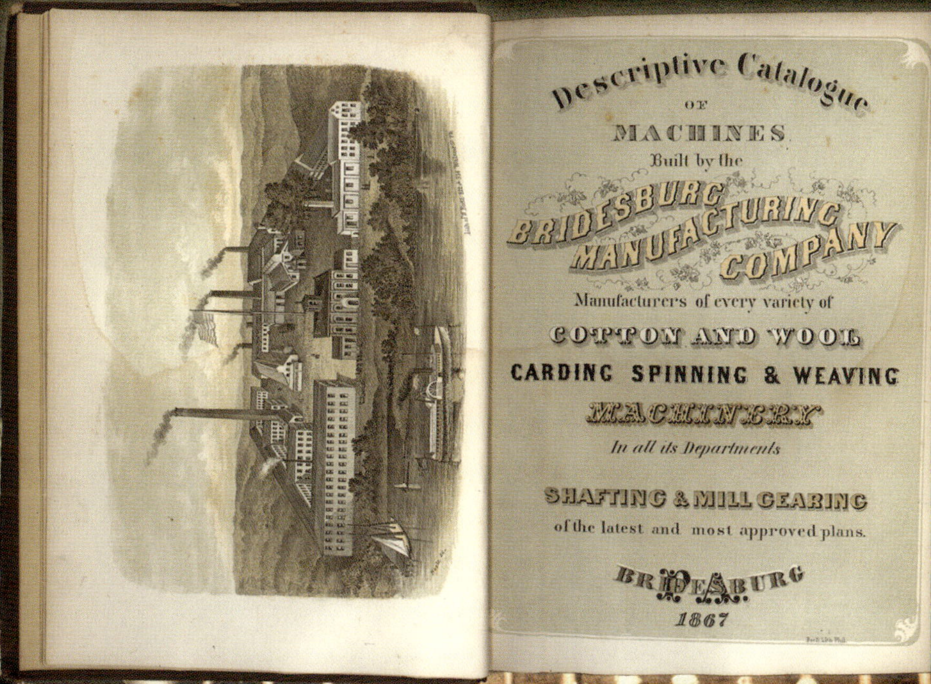

18. Background, Jenks jack detail, showing jackspool and spindles. The jack is stamped:

PAT^{D} BY JOHN GOULDING, DEC.5TH.1826.
RE-ISSUED JULY 29TH.1836.
EXTENDED AUG.30TH.1862
N^{O} 265

Photo by Jet Lowe, 1978 HAER (97).

19 and 20. Above and right, two ring-frame ply-twisters took single strands of yarn from the spinners and plied/twisted them together into wefts for the weavers' looms or for two-, three-, or four-ply yarn. The spinner measured the yarn to prescribed lengths. DNR, WMSHS.

21. Below, reel wrapped with threads from the creel stand in the warping mill. DNR, WMSHS, 2004.

22. Above, carding machines were a danger to clothing and fingers. Notice the lack of guards on the machinery. DNR, WMSHS.

23. Left, holes in wall next to looms were caused by runaway shuttles—one of the dangers of being a weaver. DNR, WMSHS.

24. Below, this hole in the floor was caused by the repetitive motion of the machine operator pumping the foot-powered hanker. After 1883, an automated hanker, probably the last machine purchased for the mill, was installed. DNR, WMSHS.

25. *Above, the natural light from the windows shows off the brightly-dyed wool made into yarn on the spinning jack.* DNR, WMSHS.

26 and 27. *The current setup at Watkins Mill has sixteen looms although Watkins initially projected installing thirty-six. One weaver could manage three plain or two drop box looms simultaneously, while a broad loom or fancy loom (left and below) required its own operator. Fancy looms, such as this Stafford loom, could weave complicated multicolor patterns.* DNR, WMSHS.

28. Right, one of the darkest corners of the building was at the first floor fulling mill. Photo by Jet Lowe, 1978 HAER (96).

29. Below, because the mill relied on natural lighting, the building had many windows; but the windows were kept closed to maintain the high heat and humidity needed to process wool. DNR, WMSHS.

30. Left, in the 1950s, the church was hit by severe storms. WMA began restoration in the 1960s. DNR completed it in the 1980s. DNR, WMSHS.

31. Right, rebuilt Mount Vernon Missionary Baptist Church, c. 1981. DNR, WMSHS.

32. Below, inside of the restored church. DNR, WMSHS.

33. Right, inside of the Franklin Academy, 1981. DNR, WMSHS.

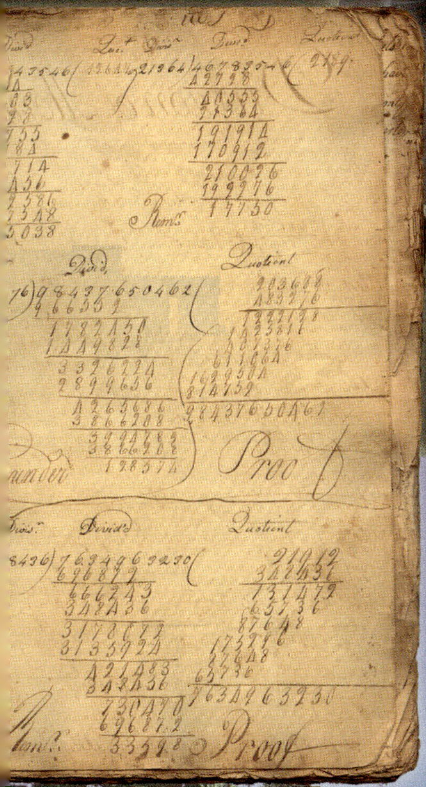

34. Left, Waltus Watkins's dedication to education is seen in his keeping of this ciphering book used between 1788 and 1793 by his father, Benjamin Watkins. DNR, WMSHS.

35. Below, octagonal Franklin Academy, built in 1856. Photo by Jet Lowe, 1978 HAER (103).

36. Right, the curving staircase in Waltus's and Mary Ann's Bethany home. DNR, WMSHS.

37. Left, while not original to the house, this piano is of a style similar to the one purchased by Watkins in 1870. DNR, WMSHS.

38. Right, the parlor of Bethany was enlivened when Waltus purchased colorful carpet (here reproduced) for Mary Ann. DNR, WMSHS.

39. The Bethany house on a bright clear winter day. DNR. WMSHS.

40. Below, the smokehouse is one of many buildings the Watkins family used to preserve food. DNR, WMSHS.

41. Below, the Bethany household was supported by the outdoor kitchen, smokehouse, and kitchen gardens, all conveniently located at the rear of the house. DNR, WMSHS.

42. Right, volunteers talk with visitors in the 1800s gardens at Watkins Mill Historic Site. DNR, WMSHS.

43. Below, the DNR assumed responsibility for the site in 1964. The WMA continued its guided tours until replaced by the living history program and tours by state employees. DNR, WMSHS.

7

Considerable Knowledge and Mechanical Ingenuity

For nineteenth-century Americans, who spent much of their working time outdoors and whose buildings were poorly heated, wool was a nearly ideal material for clothing, blankets, and other everyday needs. It is the peculiar nature of raw wool fiber that gives woolen cloth its unique properties and also dictates how wool is processed. Wool fibers are covered with microscopic scales; upon agitation, these catch on the scales of other fibers and cause the strands to become entangled in a process called felting. The resulting dense mesh of fibers will not ravel, is water resistant, and insulates the wearer against both hot and cold air. Wool fiber is highly elastic and will return to its original shape after stretching. Moreover, wool is hygroscopic, meaning it will retain a great deal of water without feeling moist on the surface.

Producing finished woolen fabrics from raw wool involved a number of major steps, each with its own machines and trained workers. As a full-production woolen factory, Watkins Mill was equipped to handle all of these.[1] Production required moving the wool from place to place in the four-story factory as each step was completed. Work areas were not necessarily arranged for greatest efficiency; rather, machinery was placed wherever it could be fitted onto the line shafts of the power train. There was much hustling and bustling about in the mill as the unskilled laborers who carried the products hurried from each machine or process to the succeeding step. Runners, including at least five children, scrambled across the floors or wearily trod the staircases to furnish the more skilled mill operatives with necessary supplies. Raw wool, baskets of bobbins and spindles, and bundles of partially finished stock were everywhere.

Raw Material

The first step in wool manufacturing is, of course, securing raw wool with which to work. As previously noted, Watkins Mill obtained its wool from local sheepherders;

from at least eight different dealers in Missouri, Chicago, and Philadelphia; and from Watkins's own flocks.

Shearing sheep was done by hand, although new equipment helped move the job along. The first patent for a mechanized sheep shear was issued in 1855. It speeded up shearing by replacing the traditional hand-operated clippers with a mechanical clipper, similar to the ones used by barbers today. However, the shearer still had to hold and position the sheep and guide the cutter head through the wool just as he had guided the old hand clippers. His assistant had to crank the shear hard enough to drive the oscillating blade through the matted and tangled fleece.

The wool clipped off the sheep varied in quality, depending on where it grew on the animal's body. The best wool, long and soft in fiber, came from the sides of the sheep. Fine wool also grew along the neck, while the wool on the back was somewhat poorer, as it might be sun damaged. A still lower grade grew on the shoulders and thighs. Other considerations, such as the breed of the sheep, the pasture soil where it grazed, additional feed it might receive, and the weather, could all play a role in the quality of a fleece. The best grades for Watkins's blankets and top-notch yarns came from the merino breed.

SORTING

Sorting the wool was a critical step. By touch and sight, the sorter would determine where the wool in each fleece changed character and divide it into grades or sorts. An experienced sorter, trained in a four-year apprenticeship, could grade fleece into as many as fourteen sorts, though four grades was the norm for American factories and probably the number used at Watkins Mill. The different types of wool would be employed singly or in combination to produce different kinds and grades of cloth.

At Watkins Mill, the sorter was stationed at a first-floor bench or, later, within the woolshed to the west of the factory building. In good weather, he might work on the outside platform on the south side of the mill. Clipping the fleece apart as he went, the sorter divided the wool into piles and removed as much foreign matter as time allowed. Generally, "foreign matter" meant burrs, clumps of dirt, and the like, but upon occasion a sorter might discover a large rock, rope, or some other heavy item that an unscrupulous vendor had added to increase the weight of the wool sack and therefore its price.

CLEANING

The first machine applied in the woolen manufacturing process was the willower. A set of large metal teeth revolved past each other, knocking out dirt, dung, burrs, and other impurities, and opening up the

Left, duster/willower, A. Jenks, Bridesburg [Pennsylvania], c. 1850. Photo by Jet Lowe, 1978 (1).

Center, left to right, main drive pulley, skein dyer, cloth dyer and washer, fulling hammer; first floor, southeast corner. Photo by Jet Lowe, 1978 (55).

Right, picker, A. Jenks & Son, c. 1865, feed drive. Photo by Jet Lowe, 1978 (3).

closely packed fibers of the fleece as well. Willowing derives its name from the willow sticks formerly used in beating fleeces by hand to open the fibers and remove the filth. Rotary willowers, based on the rotary principle applied to machinery by British innovators in the 1700s, replaced this hand labor, greatly speeding up the process. The willower used by Watkins is an older style that was probably purchased secondhand and may have been quite aged in 1861. Originally the willower was located on the ground floor, in the wool storage and sorting area on the north side of the building. Sometime in the 1860s, the gristmill was moved into this area, after which willowing took place in the second-floor picker room.

Once willowed, the fleece was carried to the first-floor dye room for scouring (washing). This procedure removed the dirt and yolk that together could account for one-third to one-half of a fleece's weight. Yolk, which gives wool its natural creamy color, is a coating of wool grease, or lanolin, combined with a watery secretion from the sweat glands, called suint. Fabric made from wool with the natural oils and lanolin left in is highly water resistant, but the oils also resist dyes. Less than a third of the wool processed at Watkins Mill was worked "in the grease."

The mill employed a simple cleansing process involving alternating rinses in a mixture of alkali soap and warm water, followed by repeated fresh water rinses in other large tubs. The scourer not only had to tolerate the stench of his solutions, he also had to pay attention to temperature and acidity and be careful not to agitate the wool too much and cause it to felt. The washed wool was laid to dry on the platform outside the south wall of the factory. In a given year at Watkins Mill, as much as forty thousand pounds of wool had to be scoured. (The wool might be dyed at this point, or later as yarn or as woven fabric. Dyeing is covered below, in the section on finishing.)

Picking

The clean dry wool was next taken to the picker. The role of this machine, with its slapping power belts, revolving power pulleys, and iron teeth whirling at over nine hundred revolutions per minute, was to pull the clumps of wool apart and loosen the fibers. The picker blew the mixed and loosened wool through a small opening into the roomlike picker bin. Refined lard oil was applied both before and after picking, which not only rendered the wool more manageable but also made picking one of the most dangerous stations in the factory. If not processed right away, the oily piles of wool were prone to spontaneous combustion. There was also a very real possibility of fire if the picker's iron teeth struck a spark on a stone or piece of wire caught in the fleece. Many a mill burned when the fiber-filled atmosphere of the picker room was ignited by a spark. Large mills usually had a separate picker house to reduce the danger—in fact, major insurance companies would not insure a mill unless it had a picker house—but smaller establishments like Watkins Mill often housed the picker in the factory itself. Whether this was due to the cost of erecting another building or to ignorance of the danger on the part of the small mill owner is unclear.

The picker machine was also used to blend sorted lots of wool to the right mix for the particular fabric. If different-colored wools were to be mixed, workers such as Doctor F. Marley ("Doctor" was his given name, not a title) spread the wool on the floor of the picker room. The wool in the predominant color was evenly arranged into a rectangle on an open section of the floor. The next color was spread in an even layer on top of the first color, and so on, until all the colors were evenly distributed. An assistant carefully removed a section of the stacked wool, and the wool picker spread it on the apron feeding his machine. The colors were thoroughly and evenly mixed by the picker machine. One of the most common mixtures was black wool and white wool blended together to form "sheep's gray." Because it could make use of white wool that did not have the best color, sheep's gray was less expensive than other fabrics of the same type. Wool jeans and other work fabrics, as well as knitting yarns, were often made of sheep's gray. Blue mix and several shades of mixed browns were also popular. One notable bolt of jeans (recorded in the Watkins Mill weaving book) was made of purple wool with yellow and white flecks mixed in.

When the wool covering the floor of the bin was about a foot thick, more oil was added. The wool was tossed and mixed with pitchforks to spread the oil evenly, then sent up to the third floor for carding.

CARDING

Carding—the process of smoothing and blending the wool fibers in preparation for spinning—was of prime importance, as all the succeeding steps in production were keyed to the uniformity of the carded wool. In colonial times, this step was done by hand, by children or adults using pairs of wire-covered boards called cards to smooth out clumps of wool. In the industrial era, skilled machine tenders (carders) fed the wool through a set of three carding engines. The cards, as the machines were called, consisted of a main cylinder (a large central drum about 3 feet in diameter) surrounded by a series of smaller cylinders: fourteen of six different sizes on the first two machines and twelve on the third. These smaller cylinders were variously labeled the lickerin, workers, strippers, fancy, and doffer. Cards worked in sets of three, known as the first and second breaker and the finisher.

The first and second breakers differed only in the way the wool was fed into them. The first breaker had a wide rotating belt of wooden slats known as an endless apron that carried the wool into the front of

Right, first breaker production card, A. Jenks & Son, with drive belt from floor below, c. 1865. Photo by Jet Lowe, 1978 (10).

Left, front view of baller. Photo by Jet Lowe, 1978 (13).

Below, overview of baller. Photo by Jet Lowe, 1978 (14).

Above, wood and iron creel feeding second breaker card. Photo by Jet Lowe, 1978 (16).

Left, cards; view looking from the southwest. Photo by Jet Lowe, 1978 (6).

Below, card, A. Jenks & Son, roll-drum and shell delivery, c. 1865; view looking southwest. Photo by Jet Lowe, 1978 (7).

the machine to the lickerin. As the raw wool was pulled into the card by the lickerin, the main cylinder picked it up and carried it past the six workers, which untangled and separated the fibers, removing knots and smoothing out the wool to form a filmy, weblike blanket.[2] The stripper, rotating in the opposite direction, pulled the wool off the adjacent worker and deposited it back on the main cylinder, to be picked up by the next worker cylinder. The six sets of workers and strippers were set gradually closer together so that each set untangled and smoothed the wool a little more. After passing through all six pairs of workers and strippers, the wool was lifted off the main cylinder by the fancy, which had longer teeth and moved slightly faster. Brushing the fine mass of wool off the main cylinder and lifting it away, the fancy transferred the wool onto the doffer. There it was pushed off by an oscillating wooden bar called the doffer comb. When carding started, the wool would begin to bunch up on top of the doffer comb. The carder would take one corner of the sheet of wool and feed it into a side-draw condenser mounted on one side of the card. A rotating trumpet-shaped tube on the condenser reduced the blanket-like mass of carded wool into a lightly twisted strand about an inch and a half in diameter. This roving, or sliver, was then wound onto a foot-long wooden bobbin. When the bobbin was full, it would be replaced and set into a rack, or creel, on the front of the second breaker.

The condenser, patented in 1824 by John Goulding, was a very important development in American carding. Prior to Goulding's invention, carding machines could only produce pieces of roving as long as the width of the carding machine (about 18 to 24 inches). A second machine, called a slubbing billy, was needed to connect these short lengths together to form a roving long enough to be spun into yarn by machine. The Goulding condensers produced continuous rovings that didn't have to be pieced.

When thirty-six bobbins had been filled and set into the bobbin creel, the ends of the rovings were fed into finger holes in a wooden board on the front of second breaker so that the wires on the lickerin could pick them up. The second breaker worked in the same manner as the first breaker, mixing the wool more thoroughly and evening out any thick or thin areas. The finisher continued the process, but here the wool was doffed differently. Instead of a single doffer cylinder with a solid covering of card clothing, the finisher had a pair of ring doffers, set one above the other. These featured narrow rings of card clothing separated by plain leather strips of the same width. The rings of card clothing alternated on the two ring doffer cylinders so that the first one pulled narrow bands of wool off the fancy, leaving the wool between those strips to be picked up by the second doffer. (The number of rings on doffers varied; Watkins's cards had eighteen on each doffer.) Because of the space the doffers took up, the finisher card had only four sets of workers and strippers in contrast to the breakers' six.

As the ring doffers turned, the bands of carded wool were fed into a rub roll condenser at the end of the finisher. This condenser consisted of two banks of rubber-covered metal cylinders. The rub rollers in each bank were set into two rows, one above the other and parallel to the card's doffer cylinders. The rubbers moved from side to side as they rotated, condensing the strips of wool into very slightly twisted strands about an eighth of an inch in diameter. These strands, or ends, were then wound side-by-side onto pairs of jackspools on a spool stand just in front of the condenser. When full, the spools were set onto tall racks attached to the support posts running down the center of the third floor.

At Watkins Mill, one set of cards was used for making batting. A revolving drum the same width as the card was placed immediately behind each of the two breakers, which operated independently of each other in this process. The sheet of carded wool was fed from the doffer comb straight onto the batting drum; that is, it did not go through the condenser, nor did it pass through a second breaker. When the wool wound around the drum had reached the right thickness, it pushed up a lever that rang a bell. The carder would stop the machine, then slice the batting on the drum so that it could be peeled away in one rectangular piece. The two breakers were different widths and produced batting for both single- and double-bed quilts and comforters. The batting card set could also be used for roving production when needed; the carder would simply feed the carded wool from the doffer comb into the condenser instead of onto the batting drum.

At one end of the carding department was a small single card used to process wool for home spinners. Instead of a condenser, its doffer fed the wool onto a wooden cylinder (shell) with grooves cut straight across it. As the mass of carded wool was fed between the wooden cylinder and a close-set curved wooden piece called a breast, it was rolled up into rolags (lengths of roving about 1 inch in diameter and 18 inches long). When a number of rolags had dropped into the bin behind the wooden cylinder, the carder would twist them into a loose ball known as a roll. The spinster would open the ball and spin the rolags, working each one onto the end of the last. These short lengths were easier for a hand spinner to manage than continuous rovings, which tended to get tangled and dirty as the spinner tried to work with them. Tradition was probably another factor in the production of rolags, as people were used to spinning the short rovings produced by hand carding.

It took two men to run a set of cards, with a card boss to oversee the department and keep the machines adjusted. One man weighed the wool on a scale mounted on the side of the first breaker (card) and spread it evenly on the apron in a loose W shape. This allowed the wool to enter the first card evenly, preventing thick or thin areas in the mass of carded fibers and insuring that the yarn spun from the wool would be uniform. Although this was a simple task, it required constant attention. Putting too much or too little on the apron would result in thick or thin spots that could not be evened out later. As he tended his machine, the carder had to be sensitive to changes in the lightly oiled raw wool at the feed, to adjustment of the cylinders of each card, and to the humidity of the workplace.

The second man removed each filled bobbin from the condensers on both the first and second breakers and set a new one in place. This was done while the machine was running, and required deft handling to smoothly transfer the growing roving onto the new bobbin. He set filled bobbins from the first breaker into the second breaker's bobbin creel and those from the second breaker into the finisher's creel. He also replaced filled jackspools on the finisher and carried them to the spool racks, ready for the spinners. (Dents in the floor between the cards and jacks indicate that Watkins's carders had different techniques in removing empty bobbins from the rack. One dropped the empties

Left, four jacks, showing the tracks, jackspools, and spindles; view looking from the east. Photo by Jet Lowe, 1978 (24).

Below, jacks; view looking from the northeast; Jenks foreground, Furbush & Gage beyond. Photo by Jet Lowe, 1978 (25).

near his feet while the other flung them behind him, hitting a small barrier wall that kept them from flying into the nearest spinning jack.[3]) By the 1860s mechanisms had been invented that transferred the rovings directly from one carding machine to another, eliminating the need for the worker who changed the bobbins. Watkins knew about these innovations but chose not to replace people with machinery.[4]

The carding engines needed a lot of attention to keep them in running order. The boss carder checked the adjustments five or six times a day. Using a set of graduated metal gauges to measure the gaps (.025–.035 inches) between cylinders, together with a sharp eye and ear, he adjusted each roller in or out, up or down, to keep it the prescribed distance from the adjacent cylinders. His skill in regulating these settings determined the thoroughness of the carding process and the uniform quality of the product.

The cards were cleaned frequently, which meant going over every cylinder with a special hand card or with a fine-toothed card comb to remove wool caught between the wires. The carders cleaned the first breaker every day, the second breaker every other day, and the finisher at least once a week but preferably every three days. In addition, the card clothing on the cylinders had to be sharpened regularly to maintain the needle-sharp wires needed to catch the individual wool fibers and pull them out of the tangle. The cylinders were set on the grinding machine two at a time, and a rotating, emery-covered grinding wheel moved slowly back and forth between them as they turned. Dressing (grinding) each pair of cylinders took two to three hours and required great skill. Variations in temperature and humidity could cause the wooden cylinders to change size during the process, the wires needed to be aligned properly, and the tip of each wire had to be slowly and gently ground off to precisely the right angle and depth. After grinding was completed, the card boss went over each cylinder and touched up any problem areas by hand. He also had to dress the main cylinder. This was too large to be removed from the carding engine, so he would use a sheet of ticking, stretched tightly over a curved frame and covered with emery, to grind the main cylinder clothing by hand.

In the hands of a skilled carder such as Watkins's Thomas McGerty, a set of card clothing could last a year. If not done properly, grinding had to be repeated, which shortened the life of the card clothing. Replacing the clothing was costly in both money and lost productivity; it took quite a bit of time for the card boss to strain on a new strip of clothing so that it spiraled tightly and smoothly over the entire surface of the roller, and the machine would of course be out of operation in the meantime.

Carders were essential craftsmen whose skills came from apprenticeship training followed by years of practice. As noted by Laurence Gross, a successful carder was conversant with all aspects of his trade, fully understanding the functions of the different cylinders and the proper operation of every part: "The carder had to encompass a variety of skills, none of overwhelming difficulty or beyond the scope of learning on the job but each an advance over the skill of hand carding—and cumulatively important. He had to be a generalist, utilizing a variety of tools and skills to keep the machine operating successfully." Watkins's own experience as an apprentice in Kentucky mills afforded him a respect for the talents and methods of the carders he employed. In 1867, he had on staff six carders and a card boss, which would be a full carding crew for a factory the size of Watkins Mill. The carders were Ab James, James Feboe, James Fitzgerald, Ben Holman, William McIntire, and W. K. McKnight; the card boss may have been Thomas McGerty, who was so listed in the 1870 census. A newspaper advertisement printed in 1864 asserted that a skilled Englishman was in charge of carding at Watkins's new mill; McGerty was actually Irish but may have trained in English mills.[5]

These men labored amidst constant danger from the sharp, bent carding-cloth wires on the revolving drums as well as from the exposed belts and pulleys. In this congested area, workers operated ponderous and threatening machinery. Hardly a carder escaped disfigurement or disability. Many, if not most, carders suffered serious hand and arm injuries, or worse.[6]

Starting around 1800, woolen carding had emerged as a new commercial enterprise in rural America. At first wool carders melded into the existing system of preindustrial production, their services bridging the gap between providers of fleece and hearthside spinners. Only later did industrial capitalists such as Waltus Watkins merge carding operations with other mechanized processes in wool production. Even then customers could contract carding as a separate service, dropping off their own wool and picking it up two or three days later. (A few chose to have the wool carded while they waited, perhaps taking advantage of the opportunity for social interaction with the family.) Customers were expected to provide the amount of oil needed to process their wool; compliance was encouraged by the high fees charged if the mill's oil had to be substituted. The aim was to avoid depleting the mill's supply, which had to be ordered and shipped by rail from St. Louis or Chicago. Occasionally wool brought in for custom carding was rejected as being too full of burrs or trash that would damage the card clothing.

Watkins's original scale and intent of operations determined that he install two sets of cards (31-inch and 48-inch models from the noted Pennsylvania firm of Alfred Jenks and Sons and a 24-inch Jenks model obsolete since the 1820s), but business quickly grew to the point where the third set was added. The ring-doffer condensers on Watkins's finishing cards were five- and seven-rub rollers. As the five-rub type was outmoded soon after the Civil War, the other probably handled the bulk of the carding in the factory's heyday. Waltus purchased his last set of cards in 1867 for $1,250 (most likely with profits from wartime sales); apparently he and his son John saw little reason later to update the equipment. For a number of years, a ready supply of inexpensive labor allowed the Watkinses to maintain this stance, but eventually their failure to invest in new machinery would catch up to them.[7]

Spinning

The next step after carding was spinning, in which the rovings with their loosely connected fibers were simultaneously twisted and stretched to form yarn. The scales on the wool fibers caused them to cling together, helping the roving to draft, or stretch out, without pulling apart. In the handicraft era, girls, women, and even elderly men (all known as spinsters) used walking, small, or great spinning wheels to make woolen yarn; even young children could twist wool on small drop-spindles. Starting about 1764, wooden spinning jennies allowed for greater production of spun yarn, either at home or in early factories. However, as noted in a history produced by the Museum of American Textile History, "the Spinning Jenny suffered two major disadvantages from the standpoint of the woolen manufacturers. It was hand powered and hence limited in the number of spindles which could be conveniently turned; nor could the Jenny produce yarn of truly uniform size, because its operator could not duplicate each of his motions with absolute precision. Without uniformly spun yarn, it was difficult to weave cloth of high quality." The spinning jack, developed in 1810 and widely used throughout America by the 1840s, overcame both problems, resulting in production of one-third more yarn from the same amount of carded wool. Spinning jacks, also known as hand-jacks, are sometimes referred to as semiautomatic because the machine did only part of the work by itself; the rest was completed by the jack spinner who operated the machine. By the 1870s, fully automatic spinning machines called mules were in operation. The more finely regulated mule, with its four hundred spindles, could spin each thread with uniform thickness, twist, and draft. (Even so, within the woolen industry, it was believed that hand-operated spinning produced better quality yarn than mules.) Since the mule was automatic, it did not require a great degree of skill to operate it; by the end of the century, many mills employed children to run their mules in order to reduce labor costs.[8]

Each spinning jack consisted of a long fixed headstock with a low, movable carriage parallel and in front of it. This carriage moved in and out along a set of floor-mounted metal tracks. The elegant machine operated in two stages, the first powered by the mill's power system, the second by hand. In the first operation, large spool drums on the headstock turned the twelve jackspools set along the top of the machine. The rovings from each jackspool passed through small feed rollers to the 216 spindles on the carriage. The spinner would place a wooden spindle over each of the metal spindles and wrap one roving strand, or end, onto it. Once all 216 ends were secured to a spindle, he would start the machine.

As the carriage moved away from the headstock, the spindles rotated rapidly. Since each spindle leaned slightly backwards towards the headstock, the strand of roving would keep slipping off its tip and wrapping around itself, twisting into yarn. The carriage moved outwards about 8 feet, gently unrolling, drafting, and spinning all 216 rovings as it went. When the carriage reached the end of its tracks, it stopped moving but continued to spin for a little while. The jack spinner could adjust the speed of the carriage and the amount of this final spinning time to suit the yarn he was making. If the machine moved slowly with lots of twist, the yarn would be thin, tight, and strong; if it moved quickly with less twist, the strands of yarn would be thicker and softer. Thirty graduated speed adjustments were possible on the two older jacks, and sixty-four on the newer pair at Watkins Mill. A clock counter, mounted on top of the headstock just in front of the spinner, tracked the number of draws and tallied the number of yards spun.

Once the jack stopped spinning, it came out of gear. The jack spinner then moved the carriage back towards the headstock by hand, at the same time wrapping the spun yarn around the wooden spindles. With his left hand he would lower the faller wire onto the strands of yarn to push them down, away from the spindle tips so that the yarn could wrap around the spindles. At the same time he turned a hand wheel mounted on the front of the carriage with his right hand, rotating the spindles in the opposite direction from the way they turned during spinning. This caused the yarn to wrap around the spindles as the carriage moved in, forming a cone called a cop on each spindle. When the carriage was all the way back against the headstock, the spinner could pause to tie up any broken strands. He then pushed the carriage firmly against the headstock to activate the gears and start the next spinning pass.

Depending on the yarn being made, the carriage moved in and out from three to five times a minute. Each pass built the cops of yarn a little higher on the spindles. When the spindles were full, the spinner and his youthful assistant (known as a doffer) would remove them and set new spindles in place. In the course of a day, a jack spinner could usually fill two and a half sets of spindles at a rate of one 200-yard cop for each spindle. The total daily output (about 107,000 yards of standard-weight yarn) was roughly equivalent to what a thousand hand spinners could produce.

Spinning took a lot of skill. The jack spinner had to know his machine, how to adjust it for the correct weight of yarn, and how much to hold back or pull out on the carriage to improve the draft (the pull on the strand) as the yarn was spun.[9] He had to be able to wind the yarn onto the spindles so that all the strands were completely and evenly wound when the carriage reached the headstock, and so that they would unwind smoothly, without tangling, during the next process. The work required concentration and a lot of walking back and forth in front of the machines, and the working conditions were difficult. The spinner and doffers often worked barefoot and in lightweight clothing because of the high temperatures and humidity levels in the mill. (Yarn is stronger and more pliable in high humidity.) To maintain appropriate humidity levels—ideally, 75° Fahrenheit with 72 percent relative humidity—the Watkins spinning and carding operations were served by stoves as well as pipes from the steam boiler.

Jack spinners in Waltus's era were most often European-trained immigrants or their second-generation American sons, and were generally the highest-paid mill operatives in the woolen industry. They had a reputation for being independent and unruly, an autonomy that was symptomatic of the struggle for rule of the workplace between workers and managers. At Watkins Mill, however, they were not the highest paid, nor is there any mention of clashes between them and the managers. (Admittedly, the Watkins labor records are sparse.) Perhaps the Watkinses used extra diplomacy to ensure that these cantankerous men would not walk away from their work and idle the factory, or perhaps the most egotistical jack spinners were never recruited to Clay County in the first place. Modern interpreters tend to believe that Waltus's personalized management neutralized these restive workers.[10]

The four spinning jacks of Watkins Mill sat in two back-to-back rows on the north side of the third floor. The two on the east end were manufactured by Jenks in 1860; those on the west were acquired from Furbush and Gage in 1867–68. The machine in the northeast corner was apparently the most extensively used of the four, as evidenced by the greater wear on the floorboards before it, as well as on the jack itself. It is closest to the windows, and prior to the acquisition of the oversized Jenks jack that sits behind it, was the most accessible and easiest to work on.

The Jenks models were probably outmoded when Watkins purchased them, a common story among industrialists on the frontier, who frequently purchased secondhand equipment. Norman Crockett comments, "Although new machinery was available, midwestern woolen manufacturers tended to buy secondhand equipment whenever possible. Such equipment was not only cheaper but it effectively served the needs of producers located in the Midwest" who concentrated on coarse and medium quality textiles. Obsolete equipment also remained in use because many owners in outlying areas were ill-informed about recent innovations in their field. When industry-wide journals became available to subscribers, this hurdle could be surmounted, but capital for new equipment remained a problem for small manufacturers like Watkins.[11]

In the 1870s, for $200, Watkins could have converted these machines to self-actuation and thereby sped up production. For $400 he could have purchased an automatic jack. Watkins choose neither innovation. The partially hand-operated spinning jacks stayed, and their operatives continued to trudge miles daily in front of the twirling bobbins and twisting strands. At a time when jack spinners were being replaced by automatic spinners in eastern mills, the Watkinses and others like them (such as the Corbins in Liberty) provided a haven for these skilled workers.

WEAVING

After spinning, baskets filled with spindles of yarn were carried down to the second floor, where weaving, as well as production of knitting yarn, took place under the supervision of the loom boss. Here the walls were whitewashed to enhance visibility, as light in the mill was limited to daylight coming through the many large windows. Although lighting fixtures were available for mills, there is no indication that Watkins ever used them. The predominant source of artificial light at the time (until 1880, when Edison installed the first electric power plant) was kerosene and lard-oil lamps. Woolen mills were necessarily closed environments; in order to maintain the needed humidity and keep the wool from blowing all over the place, the windows were never opened. Because the lamps required oxygen to burn, workers could soon become disoriented and sluggish from lack of oxygen, a truly hazardous situation for those working with machinery that was dangerous under the best conditions. The lamps also increased the risk of fire. No doubt his early experience in mills had made Watkins well aware of these dangers.

In contrast to the hulking carding machines and extended spinners, the looms were the most intricate machines in the mill. Much of the time in the weaving process was spent arranging yarn for the mechanized steps. Looms create cloth by interlacing yarn in gridlike patterns. The warp threads—those running the length of the fabric and strung in the loom so that they ran toward the weaver—were arranged by hand, with several steps required to prepare them for the loom. Once they were in place in the loom, shuttles bearing the crosswise filler thread, or weft, passed repeatedly back and forth parallel to the weaver and over and under prearranged groups of warp threads. These groups of threads were controlled by harnesses, wire-filled frames that could raise and lower a specific group of warp threads. It was in the weaving room that the female workers entered the production process; among the known weavers employed at Watkins Mill, only one was a male.

Before weaving could begin, the yarn or thread had to be prepared. The warping mill consisted of the creel stand and the reel, which stood next to the staircase on the second floor. (Local carpenter Ben Vermillion may have been the one who shaped the wooden pieces, fit them together, and added decorative touches.) Spindles of warp thread, more tightly twisted than weft, were arranged in the creel by the loom dresser according to the color and pattern being set up. A section of up to 260 spindles was laid out according to the draft, then threaded through a tensioning device and secured to the large wooden warping reel. The reel, 6 yards in circumference, was rotated until the threads were wrapped in a narrow band around it. The process was repeated until enough threads—anywhere from 360 to 3,080—were wound onto the reel to make a warp of the proper width. A counter on the reel kept track of how many yards had been wound on so that every section was the required length. Then the warp, which could measure from 30 to 109 inches in width, was tied onto loom beams of the appropriate length to fit either a narrow or a broad loom. The reel was then put

Warping mill showing the reel being wrapped from threads in the creel stand, second floor from west. Photo by Jet Lowe, 1978 (41).

into reverse; the warp beam was rotated until the warp was wound on it. This process created a smooth warp with all the threads evenly spaced and under even tension, important because loose warp threads would not feed properly on the loom. Watkins's warps were anywhere from 75 to 400 yards in length; the majority were 175 to 200 yards.

The women would then carefully and patiently draw in and arrange wool or cotton threads to create the warp. Each of the harness wires, or heddles, had an eye; drawing in consisted of pulling a separate warp thread through certain of the eyes using a long thin hook. The pattern of the cloth determined which heddle in which harness (or harnesses—some patterns required warp threads to run through more than one harness) each thread was to be drawn through. For example, in a simple four-harness twill, the first thread would be drawn through the first heddle in the first harness, the second thread through the first wire in the second harness, and so on for the third and fourth. The fifth thread would be drawn through the second wire in the first harness and the rest of the sequence repeated. This would be done over and over until all the threads were set up in the repeating one-through-four pattern. This produced a diagonal weave like the one found in blue jeans. (Many, if not most, of the fabrics Watkins made were twill weaves.) However, some fabrics such as cassimere had elaborate woven-in patterns requiring the warp to be fed through the harnesses in complex patterns. Simple or complex, each thread had to be in exactly the correct heddle in the right harness or a price-reducing flaw would be woven into the fabric. Experienced weavers (such as Ann Reid at Watkins Mill) could draw in two or three warps in a day.

If a new fabric was to be woven in the same pattern as the previous piece made on the loom, the new warp threads could be tied to the old ones and the new warp pulled through the harnesses without having to be drawn in. Although this took less skill, it did require a great deal of care and patience since every thread had to be tied on in the proper sequence.

Once the loom was dressed, weaving could begin. The harnesses raised and lowered the warp threads on alternating passes of the shuttle; after each pass, the weft thread was pushed snugly into the "web" of the woven cloth using a long comblike bar called a beater, thus straightening and tightening the weave. In Waltus's era, mechanization affected each of these components. First, rather than the shuttle being thrown by hand, it was propelled back and forth along a track at the front of the loom by wooden picker sticks activated by the power train. At Watkins Mill, the looms were set for 88 to 130 picks per minute, or up to 65 passages of the shuttle in each direction. Second, the openings between the warp threads, known as the sheds, were controlled mechanically; the harnesses were no longer raised and lowered by a weaver operating foot treadles, but rather by cams or levers (jackknives) on the loom. Finally, the beater bar was pushed by mechanical power rather than pulled by weavers' muscles.

Wool weaving had been particularly resistant to the mechanical inroads of the Industrial Revolution. Woolen yarn is more fragile than cotton and could be easily torn apart by early-nineteenth-century looms. By the 1830s, American wool manufacturers had adapted by using cotton warp and wool weft to produce satinet, a comparatively cheap and popular textile. In 1837, British emigrant William Crompton patented a loom for the production of fancy cotton goods; this fancy loom employed changeable chain-driven pattern mechanisms that were, in effect, mechanical computers controlling the loom's warp and weft threads to weave complicated patterns. After Crompton adapted his fancy loom for woolen production in 1840, fancy looms quickly became the standard for woolen manufacturing. Fancy looms had three distinct advantages: they could weave complicated multicolored patterns; the harnesses, and hence the patterns, could be easily changed; and most importantly, they were gentle enough to accommodate woolen warp.

The Watkins Mill weavers worked with a number of looms of different kinds: plain, drop box, broad, and fancy. Basic weaving could be learned within a week's training and handled even by young girls. (Mandy Dagley moved from twister to weaver soon after her twelfth birthday.) Yet it took experience and skill to stop and restart a loom without leaving a flaw in the fabric. The main tasks for weavers working

Above, Alfred Jenks broadlooms, c. 1850; second floor northeast. Photo by Jet Lowe, 1978 (45).

Left, Stafford loom, with R. W. Andrews patented pattern device, c. 1868; view looking from the front right. Photo by Jet Lowe, 1978 (48).

M. A. Furbush & Son broadloom with 1863 box motion; second floor north. Photo by Jet Lowe, 1978 (53).

power looms were to keep the shuttles filled with weft thread and to guard against broken warp or weft threads. At the height of its production, Watkins Mill employed between ten and sixteen weavers. Watkins sought to stretch out his skilled labor by positioning some looms so that one weaver could manage three plain or two drop box looms simultaneously. (Thus these looms faced each other.) A single broad or fancy loom required more attention from the operator. The placement of the looms, largely dictated by power train arrangements, was not always conducive to efficiency.[12]

Watkins had planned originally for thirty-six looms; he initially installed two dozen—a large operation by midwestern standards, although eastern mill companies would consider his factory only medium-sized. He added three more broad looms in 1863 as the Civil War market boomed, and three fancy looms after the war. The former were equipped with three shuttle boxes that made it easy to weave three weft colors into the fabric; this resulted in products such as Watkins's signature striped blanket, which had groups of red and yellow stripes on a white ground. The fancy looms, manufactured by Furbush and Gage, were capable of weaving the most intricate geometric patterns. (Materials so woven were called "fancy goods," hence the name of the loom.) The fancies at Watkins Mill held up to twenty-three harnesses and could use five different colors of weft. The loom boss or weaver, by

Left, Furbush & Gage fancy loom head motion, c. 1863, shows heddles and harnesses. Photo by Jet Lowe, 1978 (50).

Right, weave room, second floor, view looking from the northeast. Photo by Jet Lowe, 1978 (46).

Right, Bridesburg Manufacturing Company twister, c. 1860s, second floor northwest. Photo by Jet Lowe, 1978 (33).

Below, skeiner, maker unknown, second floor west. Photo by Jet Lowe, 1978 (34).

adjusting the chain-driven, cam-activated harness controls, could create a design per instructions from a trade journal. Intricate herringbones and twills became favorites locally; other designs were also available. Samples of complex patterned cassimeres woven at Watkins Mill are still extant at the mill site. The cloth woven on plain and drop box looms measured 30–48 inches wide, while broad looms yielded widths of 48–110 inches; the length of the fabric was determined by the length of the warp.

The various looms illustrate the progression of mechanical innovations during the factory's lifetime and also testify to the dynamism of the factory. Keeping so many different machines running presented a considerable challenge for the mill staff. Gross has commented, "This weave room represented a loom-fixers' nightmare.... Two types of broad looms, three of plain, and two of fancy, for a total of sixteen different machines, would have disheartened the most able mechanic." In the mill can be

Weavers were paid on a piecework system, which necessitated production monitoring and recording on a daily basis. The result is that records of the weavers' productivity at Watkins Mill are more complete than documentation of most other operations at the mill. Especially rich in detail are extant warping instructions and formulas. These were intended to ensure quality control and uniformity in production of popular materials, no matter when the weaving was done and by whom. The records show that the Watkins weavers each wove about 25 yards per loom per day. In comparison, a good hand-weaver could produce 9 yards of fabric per day.

seen innumerable instances where someone, in the name of maintenance or with the goal of innovation, tinkered with the apparatus. Such mechanical adaptations might have been the work of Waltus Watkins or of Jesse Walker, who was employed as engineer and machinist from 1862 to 1877. In Gross's opinion, "These alterations point toward considerable knowledge and mechanical ingenuity on the part of the person responsible for them. Despite the mill's location and limited operation, it attracted able people to its workforce."[13]

In the mill's earliest years, the looms rarely worked five consecutive days; there was little sense or need for the factory to be in full production. Between 1866 and 1870, work became more regular; it increased through the 1870s and then began to decline in the early 1880s. From 1886 onward, weaving was at a standstill but for an occasional blanket; the factory concentrated instead on yarn production.

The northwest corner of the weave room housed two ring-framed ply-twisters, used for winding strands of yarn together to make the heavier multiple-ply yarn needed for knitting and some weaving.[14] A twister (such as 11-year-old Mandy Dagley, hired in 1866) made sure that either two or three strands of yarn were fed under a ring-shaped traveler and secured to one of the wooden bobbins mounted on the front of the ply-twister. The ring carried the strands around a housing that moved up and down around the bobbin, causing them to twist around each other. The two- or three-ply yarn was wound around the skeiner, then tied off to form 580-yard loops, or skeins. (Yarn was sold by weight; since all skeins were the same length, a three-ply skein weighed more, and cost more, than a two-ply skein.) Each skein was then twisted into a hank, a tight spiral that kept skeins from getting tangled and could be easily opened at home for knitting. A worker looped the skein over a hook on the foot-powered hanker, located in the middle of the room, and held one end securely while the hook rotated. Then, gripping the twisted skein in the middle with one or two fingers, she slipped the held end over the hook and allowed the hanker to revolve back to its original position. This caused the twisted skein to wrap around itself, forming a compact hank. Two to four dozen hanks were compressed in the nearby bundling press—a device fabricated by Watkins's blacksmith—and tied into compact bundles ready to be shipped to customers. At least two people were needed to handle all the twisting functions. Most likely the twister took care of plying and skeining while another hand such as the warp dresser did the hanking and bundling.

INSPECTING

Following completion of weaving, the fabric was removed from the loom and brought to inspection tables. The general practice in the woolen trade was to assess fines to weavers if inspection revealed subpar workmanship. First, burlers spread the bolt of cloth over a slant-topped burling table and checked its entire length for loose ends, snagged threads, knots (called burls), or other imperfections. Loose ends were clipped off, and knots were pulled up on the back of the fabric, to be removed by the cloth shear during the finishing process. The idea was to make all necessary repairs so they would be hidden on one side of the cloth. (Hence all woolen cloth has a right side and a wrong side to the practiced eye.) Next, in a process known as perching, the bolt was hung over two overhead perch rollers suspended from the ceiling. The inspector stood between the perches and pulled the bolt down so light from a window shone through it, looking for flaws the burler couldn't feel. After inspecting, the woven fabric was cut into 50-yard lengths and carried to the first floor for a series of procedures known as finishing.

FINISHING

Fulling

To make it heavier and denser, woolen cloth goes through a process of washing, shrinking, compacting, and felting that is known as fulling. Like carders, fullers had provided their services to weavers of homespun long before the Industrial Revolution. In early fulling mills, cloth was compacted by beating it for hours with huge wooden trip hammers; by the 1840s, these had been replaced by rotary fulling mills. Watkins Mill had two large rotary fulling mills, located on the ground floor; only one remains at the site today.

Fulling requires three components: lubrication, agitation, and heat. Soap provided the lubrication and controlled the degree of heat generated. Too much and the cloth did not get hot enough to full properly; too little and the heat damaged the cloth. The soap could be applied laboriously by hand or quickly and more evenly by a soaping machine. No soaper remains in the mill, but Watkins was well aware of such machines, and the overhead power pulleys suggest that one sat beside the remaining fulling mill.

Agitation was provided by the fulling mill. After soaping, the edges of the cloth were sewn together to form a long tube, which was

Left, fulling mill, first floor; side view looking from southeast. Photo by Jet Lowe, 1978 (68).

Below, fulling mill, front view, first floor. Photo by Jet Lowe, 1978 (57).

threaded through the large wooden rollers and flaps of the fulling mill. The ends were then sewn together to form a continuous loop. The cloth was run through the fulling machine, usually for two and a half to three hours. Compression between the huge rollers squeezed the warp fibers together, narrowing the cloth; at the same time the weft threads were forced together by the heavy flaps. As the cloth fell past the flaps, it tended to stretch slightly from its own weight. The fuller had to orchestrate the sideways shrinkage and the lengthwise compacting and stretching to arrive at a fabric that was evenly compacted, had the desired degree of felting, and was the required width and weight. If a very heavy cloth was needed, he gradually added flocks, tiny chopped up bits of wool fiber that would felt into the cloth, increasing its weight and thickness.

The fuller had to have a high degree of skill acquired through years of training and experience. Gil Jackson was the Watkins fuller for nearly twenty years. He learned his trade in the English woolen and worsted manufacturing town of Bradford in the industrial district centered at Leeds. Perhaps propelled by the financial panic of 1857, Jackson immigrated to America in 1858; he moved to Watkins Mill in 1860.

Washing and Bleaching

After fulling, the fuller put the cloth through the washing machine four times to flush out accumulated dirt, grease, and soap. As the fabric cycled through the washer, it was repeatedly run through a cold water bath and then through large rollers that squeezed the water out. Fuller's earth, a special absorbent clay, was added to soften the cloth and remove any remaining soap.

At some time, the Watkinses constructed a brick bleach house to carbonize the wool. The purpose was not to whiten the wool—carbonizing actually gave it a yellowish tint—but rather to remove vegetable matter. Such matter did not take dyes as well as wool and would leave visible specks in the fabric if not removed. The process relied on fumes from a heated solution of sulfuric acid to burn away the unwanted matter. It could only be used with all-wool fabrics and was most applicable in fleece markets known for excessive burrs. The oral tradition associated with Watkins Mill includes testimony from one worker who remembered that the bleach house "smelled like hell was a half mile away."[15]

Dyeing

Dyeing took place at one of three points in the manufacturing process. The dye kept its color best if dyeing was done right after scouring, or "in the wool" (giving rise to the idiom for thoroughness or steadfastness). Wool might also be yarn-dyed (after being spun into yarn) or piece-dyed (after weaving, fulling, and washing). Most of the wool used in Watkins Mill was dyed in the wool. Piece-dyeing was also used to provide the range of colors customers wanted; the mill would weave bolts of undyed white cloth and then dye just the smaller yardage needed for a specific order. Fabrics like jeans and satinet that had cotton warps might have to be speck-dyed. In those cases, the dyer prepared a special speck-dyeing solution that would color any vegetable matter in the wool weft in the cloth, blending the specks into the wool.

The early-nineteenth-century dyer, like his ancient counterparts, largely relied on natural ingredients such as sumac or walnut hulls. Imported dyestuffs included cochineal, a red gained from dried insects found in Mexico and Central America; and indigo, a blue whose attributes included its rich hue (today it gives jeans their distinctive blue color). Modern European synthetic aniline dyes, developed in 1856, did not penetrate American markets until the 1870s.

Dyeing at Watkins Mill took place under the watchful eye of William Waers, who served as dyemaster from about 1865 until 1883. He had previously worked at mills in his native Ohio and in Mirabile, Missouri, where he developed the extensive knowledge of chemistry and the respect for his materials (which included poisonous chemicals, caustics, strong acids, and other toxic substances) needed to successfully and safely manage a dyehouse.

Left, ingredients stored in the first floor dye room, used by William Waers, dyemaker and bookkeeper. Photo by Jet Lowe, 1978 (58).

Below, William Waers, dyemaker and bookkeeper. DNR, WMSHS.

Like other dyemasters, Waers collected recipes for substances used in the dyeing process; he also experimented in developing some formulas of his own. His dyebook, which is extant among the mill records, contained a total of 150 recipes; these included 132 colors, mordants (used to fix or set the color), other dyeing liquors, and waterproofing for cloth and leather. The recipes date from 1824 to 1883 and were collected from at least twenty-nine different sources, primarily dyemasters at other mills, or textile manufacturers' journals. A Mr. Stibbs—probably the master dyer who supervised Waers's apprenticeship in Ohio—is credited with fifty-eight recipes. Six of the recipes date from 1824 and are credited to a Dr. Parker. Among the trade journals that Waers found appealing were the *Independent Record Supplement* and *Woolen Manufacturer.* Waers credited Watkins with ten of the recipes and listed himself as originator of eleven.[16]

The dyebook reveals that Waers was constantly experimenting with new colors or looking for better formulas for the old standbys. A number of later recipes used aniline dyes. Some recipes were only one line long (no. 34, Ruby), while others stretched an entire page (no. 103, Violet Aniline Purple). Instructions for some dyes specified use only on raw wool; others were to be used in yarn-dyeing or piece-dyeing. The

In addition to "Rules to Be Observed in Dyeing," the dyebook also contained rules for counting interest and for measuring corn in the crib or pen, indications of Waers's diverse interests. A prototype of the middle manager in modern manufacturing concerns, Waers sought opportunities to manage capital. As the Watkinses' bookkeeper, he kept the ledgers for the factory owners. In 1889, he was appointed trustee of a $1,500 fund for a local church; he agreed to pay 6.67 percent interest per year for use of the fund, with which he purchased 74 acres of land.[a]

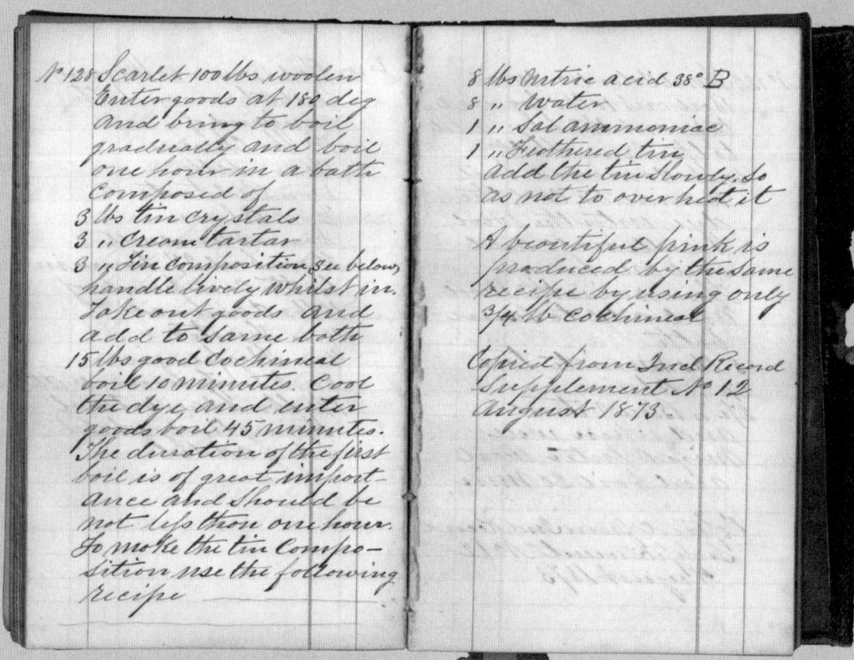

No 128 Scarlet 100 lbs woolen
Enter goods at 180 deg
and bring to boil
gradually and boil
one hour in a bath
Composed of
3 lbs tin crystals
3 " Cream tartar
8 " Tin composition see below,
handle lively whilst in.
Take out goods and
add to same bath
15 lbs good Cochineal
boil 10 minutes. Cool
the dye and enter
goods boil 45 minutes.
The duration of the first
boil is of great import-
ance and should be
not less than one hour.
To make the tin compo-
sition use the following
recipe

8 lbs Nitric acid 38° B
8 " Water
1 " Sal ammoniac
1 " Feathered tin
add the tin slowly so
as not to overheat it

A beautiful pink is
produced by the same
recipe by using only
3/4 lb cochineal

Copied from Ind Record
Supplement No 12
August 1873.

Watkins Mill dyebook open to the directions for scarlet and pink colors; tin, cochineal, and nitric acid are some of the ingredients in this formula copied from the Independent Record Supplement. DNR, WMSHS.

[a][William Waers], memorandum, 15 June 1889, in ledger of cloth samples, 267, WMSHS; Waers and Jackson Research Files, WMSHS.

dyebook includes Waers's comments on specific formulae, such as "A very good color" or "I have not tested this." Waers had a particular interest in recipes for fast (non-fading) black and for scarlet and also experimented with twelve to fifteen varieties of brown. He recorded the results on the back of old mill shipping tickets, often accompanied by glued-on samples of dyed wool.

Dyeing by William Waers was largely a hand operation. In-the-wool jobs were dyed in the dye shed; yarn-dyeing and piece-dyeing took place in the dye room. The dyeing equipment at the factory included seven large vats. Black was done in a copper kettle capable of holding four hundred gallons, but because metal affected colored dyes, wooden vats were used for those jobs. The dyer could then add alum to the mordant for a clear dye, copper sulfate for a blue or green hue, chrome for a warm shade, or iron to give a grayed tone. Waers preferred to use mineral-free rainwater rather than well water, since minerals could affect the dyes.

Final Steps

The wet and dry finishing were also overseen by William Waers. The first step after washing or dyeing was to run the bolt of wet, ropey cloth through the stretching and rolling machine to pull the wrinkles out. The cloth was then placed on the teasel gig mill, located in the finishing room at the back of the first floor. The gig mill was covered with thirty-six metal flats holding dozens of fuller's teasels. These dried flower heads from the teasel plant have stiff, hooked bristles that brushed and raised the nap, or surface fuzz, on the cloth. The fabric went through six runs, passing up and down across the teasels for a total of thirty-six brushings. The process started with used teasels; after each run, six flats were replaced with new flats filled with sharp teasels ("sharp work"). By the end of the process, all of the teasels had been replaced, and all the loose ends and other fuzz had been gradually lifted above the cloth surface. This was the end of "wet finishing."

The napped cloth was next carried up to the fourth floor of the mill to be stretched on tenter racks. Built down each side of the mill's attic, the timber tentering frames were part of the structure of the mill. Needle-sharp tenterhooks were set at three-inch intervals along the face of the top frame member. A similar row was set into a jointed board that could be raised and lowered along the base of the frame. The edges of the damp fabric were strung on the tenterhooks, and the jointed board pushed down as far as possible and pegged into place to stretch the cloth tightly between the two rows of hooks. (This is the source of another idiom, "on tenterhooks," meaning "in a state of suspense.") The bolt was brushed to lay the nap in one direction, and left to dry.

Once the cloth was dry, it was taken back to the finishing room and brushed once more to loosen the nap. It was then run through a rotary shear, which gradually trimmed the nap to the desired length. The longer the nap, the softer and fuzzier the surface of the fabric. Beaver was left with a long nap resembling fur, while flannel had a short nap. Broadcloth and similar smooth fabrics were sheared threadbare, that is, right down to the surface, removing all the nap and revealing the threads in the cloth.[17]

The sheared cloth was dampened, carefully folded into accordion pleats with press paper between all the surfaces, and placed in the big screw press between heavy iron press plates that had been heated

Left, rotary shear, first floor finishing room; view looking from front left. Photo by Jet Lowe, 1978 (66).

Below, up and down teasel gig, c. 1860s, with cloth press beyond, first floor finishing room. Photo by Jet Lowe, 1978 (60).

under the boiler. The cloth was tightly clamped and steam-pressed for twelve hours to set the nap and give the surface a smooth, lustrous appearance. The cloth was rolled into manageable bolts and was finally ready for sale.

Not all fabrics went through the entire set of steps. Jeans was a "hard finish" fabric that was ready for sale after washing. Fancy goods and fine suiting like cassimere, flannel, broadcloth, or doeskin required the final steps to give them the right body, feel ("hand"), and surface finish.

At Watkins Mill, the manufacturing process took around a week and a half from raw fleece to finished fabric. By the end of the process, the material would have passed through dozens of separate steps, most associated with complex and specialized machines operated "with considerable knowledge and mechanical ingenuity" by the mill's trained and experienced operatives. Yet individual workers no longer had the comprehensive skills to see the product through from start to finish.

Notes for Chapter 7

[1]Crockett, *Woolen Industry,* 85; *Homespun to Factory Made,* 4.

[2]The carded fibers in raw wool are not parallel, but might be caught in the middle or near one end by the teeth of the carding machines and actually lie bent across other fibers. Fibers of various lengths lie together haphazardly, with their ends scattered over the fabric surface (which is what makes wool feel scratchy). In contrast, the fibers of cotton fabrics and worsteds (which are also made with wool) are combed straight and parallel. In addition, only the longest wool fibers are used in worsteds.

[3]Gross, "Watkins Mill," 8–12; Gross, "Importance of Research," 23–25; Matthews, "Building as a Document."

[4]Gross, "Importance of Research," 23.

[5]Gross, "Wool Carding," 810; Population Schedule, Ninth U.S. Census, 1870.

[6]Gross, "Wool Carding," 817.

[7]Gross, "Watkins Mill," 8–12, 14–16; Gross, "Importance of Research," 23–25; Matthews, "Building as a Document."

[8]*Homespun to Factory Made,* 74.

[9]An industry standard was devised for yarn in which the length of yarn provided from one pound of wool determined its "count." The standard was 1,600 feet of yarn for a pound of wool. The lighter the weight of the yarn, the more lengths it would take to make a pound and thus the higher the number. For example, if it took three lengths of 1,600 feet each to weigh a pound, the yarn was labeled 3.

[10]Gross, "Importance of Research," 23; Gross, "Watkins Mill," 29–30. Only two spinsters are mentioned by name in extant records, Mike Murphy and Fred Hoffman, who appear in lists for the period 1870–72; Murphy also boarded with the Watkins family.

[11]Crockett, "Westward Movement," 114.

[12]There is some archaeological evidence that, after the 1868 purchase of additional looms, the second-floor machines were rearranged. For example, the overhead line shaft on the north side of the ceiling was repositioned at least once. Rearrangements also occurred during the period 1886–92.

[13]Gross, "Watkins Mill," 52–53, 43. Robert Gordon and Patrick Malone (*Texture of Industry,* 384) note that Watkins's looms "provide material evidence of technical solutions by anonymous artisans whose ideas were never patented."

[14]The ply-twisters are among the most prized machines in the Watkins State Historic Site collection. They are thought to be among the last surviving examples on the continent, although they were not rare a century ago. G. R. Cooper, "Watkins Mill Revisited."

[15]Goodwin, Interview.

[16]The flyleaf of the pocket-sized ledger book is inscribed "W. H. Waers, Jericho, Ohio, Sep. 9th, 1865," suggesting that it was started around that date, perhaps during a visit to Ohio. While there, Waers stopped at Charles Kappel's mill in Zoar and John Grebel's factory in Jericho. [William Waers], Watkins Mill dyebook, 1865–83, WMSHS.

[17]In preindustrial times, male artisans wielded handheld teasels and gigantic hand-operated shears. As these skilled operatives were rare in America, there had been much incentive to develop mechanical teasel gigs and shears to replace them. *Homespun to Factory Made,* 94–96.

8

None but Sober and Industrious Men Need Come Here

At Watkins Mill as well as at other emerging factories in the nineteenth century, there evolved a new system, which today is known as industrialization. Workers, often possessing crucial skills, came together in greater numbers in workplaces of greater scale than ever before. At Watkins Mill, thirty to forty operatives labored in a structure providing about 15,000 square feet of work space, where a large variety of semiautomated machinery powered by the steam engine required their attention. The new system required adjustments to traditional work habits and mores. The owner-managers determined the hours, days, and seasons of work; chose what to produce; and took control of production. Rather than relying on exchanges of services, labor was paid by wages. Managers measured outputs by profits rather than such intangibles as quality of craftsmanship.[1]

Details about the people who worked at Watkins Mill and the conditions under which they labored are scant. Sparse and scattered records such as ledgers, daybooks, correspondence, and store accounts tell something about wages, schedules, and skills, but many facets of the workplace were simply taken for granted and therefore never chronicled by managers or laborers. Diaries, which might have revealed the personal thoughts and lives of the employees, are generally lacking. How workers dealt with the thunderous clatter, whirl, chug, and clank of the power train, steam engine, and machinery; how slippery the lubricant-saturated floors were; how the high humidity affected those who monotonously walked miles daily at the spinning machines—the answers to these and other key questions were never recorded. The story must be gleaned from surviving documents and from the physical evidence of the mill itself. From these it is possible to piece together a picture of the Watkinses' relationships with, and treatment of, their employees.

PATERNALISM

The interrelationships of nineteenth-century textile mill managers and their employees were largely characterized by various forms of paternalism. In a paternalistic system, owners and managers such as Waltus Watkins treated their workers in a familial and personalized manner, rather than adopting an authoritarian or bureaucratic format. This represented a step in the transition from a traditional barter economy to the industrial-based fixed-price market economy. Historian Philip Scranton has explained, "For nearly a century, in its variations... paternalism served workers as a line of resistance against the atomization of labor and served capitalists as a strategy that married profit with the preservation of the customary duties and status of community leaders."[2]

Watkins's paternalism, as outlined in Scranton's model, "involved overlapping spheres of provision, protection, and control.... [P]aternalism evoked more than the exchange of labor for a pay-packet." Watkins provided cabins to house some of his workers, and room and board in his household for others. He also hired workers' kin, as demonstrated by the employee records for the year 1870, which show two Waerses, two Tolers, two Donahoes, three Griffins, three Taylors, and five Marleys working in the factory. Some mill workers were given work on the farm when the factory was not operating, such as during the winter when lighting was poor and water frozen. Both generalists such as Jasper Sharp and specialists such as dyemaster William Waers and carder Thomas McGerty took advantage of such arrangements (which were not entirely paternalistic, as they may have helped in retaining skilled workers). Watkins's protective role was also reflected in turning away workers known to drink alcohol or in keeping skilled workers employed rather than investing in machines to replace them. In the larger community, Watkins built the school and church, subsidized the leadership of both institutions, and monitored their agendas. By such provisions and protection, Watkins gained a measure of control in the workplace and outside it. Yet he did not impose total social control. For example, even though he did not use tobacco personally, he allowed locally grown tobacco to be sold in his store.[3]

Industrialization "involved not just a single kind of innovation but a range of new ways of living and working. And it also involved the maintenance of old ways." Watkins's paternalistic style of management was symbolic of those old ways. He did not share, in David Thelen's words, "the new order's economic imperatives of competition and growth[, which] drove an ever-widening wedge between economic activities and social and cultural traditions. The new order replaced the authority of persons with an invisible hand that created new kinds of dependence, shame, guilt, success, and failure." Watkins was more a master craftsman than a master of business, with a hands-on approach that largely proved successful in his varied endeavors.[4]

The Watkins Mill Women

Although the female weavers left few personal documents, census reports combined with Watkins's business records provide a glimpse of these women and their roles in the factory. A general ledger recorded pay of twenty women for thirty-nine months in 1863–66; the mill store kept accounts; and Watkins kept track of fabric production in a series of weaver's books, two of which survive. The ages of the women ranged from 12 to 30 years. Most (thirteen) of them worked for fewer than twelve months, six put in twelve to nineteen months, and one persisted thirty-five months. A study of the weavers observes, "Their irregular work weeks allowed them to intersperse their domestic chores at home with their gainful employment at the Mill.... [T]hey were able to straddle the transition between the household economy and the market economy by literally living in both simultaneously."[a]

Like most mill workers, the weavers endured particular occupational hazards. However, for the women, there was no possibility of working up to another post in the factory, not even loom boss. Little solace was gained from knowledge that technological and other improvements since the early part of the century had cut the weavers' workday from fourteen to ten and a half hours and had lightened considerably the physical requirements of their labors.

[a]Prude, *Coming of Industrial Order,* xii; Lentz and Wilson, "Mill Girls."

THE LABOR FORCE

To recruit the skilled textile workers needed in the factory, the Watkinses relied upon personal correspondence, word of mouth, and trade journals such as the *Woolen Mill News.* Dyers, finishers, fullers, and other craftsmen initially were recruited from European or eastern American firms. Such employees not only possessed the necessary skills and experience, but also were already accustomed to industrial culture—an advantage over local farmhands or casual day laborers, who may have had trouble adapting their traditional work habits and values to meet Watkins's expectations. Surviving documents outline some of the qualities Waltus and John sought in their workers. For example, Waltus wrote to one job-seeker, "We want a man that is qualified to run the machinery, and that is industrious, carefull, and will work to the interest of the mill and proprietors." John told another, "Men with families suit us best.... We also need some weave and card room help. None but Sober and Industrious men need come here." In return the Watkinses offered competitive wages, free housing, and, perhaps, more pleasant working conditions than those found in other factories.[5]

The applicants' replies often revealed not only the attributes they claimed to possess but also their awareness of the needs of the mill owner. Applicant James Ash wrote from Menasha, Wisconsin, in January 1884, "I am strickly temperate and not afraide of work I wood make my self usefull [in whatever position might be available]"; apparently he

The census worker who enumerated the Watkins Mill community for the 1870 census assigned the Watkins household the number 155. He listed its members as 63-year-old farmer and manufacturer Waltus Watkins with his wife, two daughters, and four sons (28-year-old John was listed as a farmer and 18-year-old Jewell as a factory worker). The Big House residents also included a diverse band of boarders: a brickmaker (age 23), a clerk (67), a superintendent (30), and three young male factory hands, two of whom were from Europe. As the census-taker moved on to domiciles 157–165, he counted "William Waters," listed as a finisher; Doctor Marley the wool picker and his two sons, who both worked in the factory; superintendent James Griffin; two weavers in the Griffin home; Gil Jackson the fuller; and Hannah and John Donahue, who both worked in the mill.[a]

[a]Population Schedule, Ninth U.S. Census, Washington Township, Clay County, MO, 1870.

offered the versatility and value system the Watkinses sought, as he was hired to begin work in May of that year. Another reply was addressed in a typical formula: "Am married man. My wife is a good weaver. I am sober, steady and industrious. Can make any kind of goods you wish...can do repairing."[6] This letter suggests the applicant may have known about specific aspects of the Watkins Mill operation, such as that Watkins favored generalists and industry veterans and that aged equipment beckoned for constant maintenance. Jobs for relatives was a frequent theme; many respondents emphasized that they would move to western Missouri to practice their trade if Watkins would also employ their female kin as weavers and their offspring in various unskilled positions. Watkins's paternalism deemphasized the contrast between farmwork and factory work and thus made industrialism appear less revolutionary to agrarian households; it "smoothed the path from field to factory; the boundaries between farm and mill were permeable."[7]

Extant agreements between managers and laborers at the mill are meager. Hiring was a personal bond symbolized by a handshake, rather than an impersonal, written contract. Perhaps representative is an 1872 notation that reads, "Engaged Mr. McLane to work in the factory at $2.50 per day if he proves satisfactory. If not we pay his board and let him go. He is to work at looms or cards or both if necessary." Another agreement noted the hiring of Jasper Sharp in 1872 as boiler fireman. If he proved satisfactory, his wages would be improved. Also his horse could be left in a nearby pasture while Sharp worked. There would be no charge "if the horse is not troublesome." A third document describing the hierarchical chain of control from proprietor through supervisor to laborer, pointed out that another generalist, Samuel Jones, would be paid $2 per day "to work at anything that we have for him to do and is morally bound by stipulations" made by the owners to one of their superintendents, Mr. Walker.[8]

Piecing together employment and census records for 1850 through the 1890s reveals staffing patterns at Watkins Mill. (See for example Appendix C, "Hierarchy of Watkins Mill Employees, 1879.") One hundred forty-three individuals have been confirmed or reported as being employed by Waltus Watkins; ninety-eight of these worked solely or predominantly in the mills. While many appear only in one year, there are some with longer service records. Skilled and supervisory personnel were more apt to remain than move on. Examples of the former include Gil Jackson (fuller), 1860–72; Jesse Walker (engineer and machinist) 1862–77; Esseks Reeves (house servant) 1864–84; William Waers (dyer and bookkeeper) 1865–83; James Rhodus (stonemason and farmhand) 1871–1905; William Reed (miller) 1870s–1888; Bill Cavender (sawyer) 1880s–1910; and four weavers (Bettie Hulls, Dorothulie Toler, Elizabeth Toler, Frances Vermillion), who all served at least a decade in the 1860s–70s. Less skilled operatives tended to move on from mill to mill every eighteen months or so, but a lack of continuity in the extant Watkins Mill records makes it impossible to determine just how transient Watkins's workforce really was. A synthesis of numerous studies of the patterns of movement in nineteenth-century America concludes, "Transiency was one visible sign of changing times.... [U]pwards of 60 percent of traced populations invariably disappear from [decennial census] records, an indication that only two in five Americans stayed in their communities for extended periods of time. These figures hide... the sizable numbers of people who moved in and out of communities within the intervals between census enumerations."[9]

Daybooks and ledgers indicate clear gender divisions in the work at the mill. Men served in supervisory positions as engineer and loom boss, as well as occupying skilled positions as bookkeeper and dyemaster. Women nearly monopolized the position of weaver and were the only workers paid piecework. All females were listed as "weavers" or "mill workers," in contrast to males being listed as "mill hands." The mystifying exception was Eliza Turner; oral tradition gives her the role of spinner, usually a man's job.

Working Conditions

The working conditions at the woolen factory can be examined in terms of both a physical aspect and a social aspect. The often grueling work, the physical environment in which that work was performed, the compensation system, and the social manipulations exerted by those in charge all revealed the ambivalent nature of paternalism as manifested at Watkins Mill.

A Dangerous Environment

Robert Gordon and Patrick Malone have noted that much work in textile mills "demanded only modest levels of skill and was both monotonous and uncomfortable to perform. A great deal of the average operative's working day involved observation of machinery and constant attention to the same process, hour after hour.... Most workers stayed on their feet all day, walking miles within a small area.... In some mills, the floors vibrated violently, and noise levels were deafening; in others, dust and loose fibers filled the air." From all indications, this description fits the Watkins operations.[10]

Physical evidence at the factory helps create an understanding of the working conditions. The internal layout of the mill, as previously noted, was contingent on the requirements of the machinery rather than on the comfort, or even the efficiency, of the operatives. Each specialized task was performed at a separate work station, but sequential stations might be scattered throughout the mill on different floors or at different places on the same floor. Worn floor boards show the monotonous miles walked by jack spinners, while pitted walls exhibit the hazards of shuttles flying off looms. A hole in the floor next to the foot-powered hanker, and the fact that an automated hanking machine was not purchased until about 1885, bear witness that for at least two decades the job of hanking tested the physical tolerance of the workers assigned to do it.

Though work in the woolen factory was probably less physically exhausting than work on the farm, it was more numbing, and also subject to unique environmental factors that jeopardized the laborers' health and comfort. In the first place, the din of the workplace, equated with that of a modern rock concert, can now only be imagined. In the weave room, for example, the clattering looms each operated at about 105 decibels. A letter from a visitor told of hearing the factory noises from two to three miles away.[11]

While workers throughout the factory had to endure the high temperatures and humidity levels, noxious and toxic air was a particular problem for certain workstations. No fresh air was permitted on the second and third floors. Fibers filled the air, especially in the eastern end of the second floor where the picker whirled. Sorters were subject to "woolsorter's disease" (anthrax), dyers handled toxic and caustic chemicals, and the odor in the bleaching area was highly sulfuric.

Low light levels taxed workers' eyesight, particularly that of the weavers, who strained to spot broken warp threads. As noted, operatives probably relied on natural light throughout the sixty-six-hour work week. There is written evidence that in 1866–67 some workers, taking advantage of lengthening daylight in June and July, may have toiled for sixteen hours each day. Of course, the longer they labored, the more prone to accident they were.[12]

Accidents could be caused by general work conditions as well as threats from specific machines. Throughout the factory, the leather belts and heavy metal gear teeth made working near any machine dangerous. Any misstep or worker negligence could be perilous. Sure footing was not the norm. Modern visitors find narrow and steep stairs hazardous, but where they now tour the second and third floors on rubberized treads, workers more than a century before slipped and slid across damp oiled boards. Lubricants from the drive train spattered over drip pans, and grease from the wool built up around carding and spinning machines. Operatives of the carding machines were prone to lose fingers (or more) in the whirling metal teeth or bent wires. In the weave room, a misfired shuttle could bruise or injure anyone in its path. Female weavers tied their hair tightly into buns lest it become entangled in belt-driven looms. (Eastern mills prohibited the wearing of jewelry and hoop skirts.) By modern standards, factory work was perilous. Injuries undoubtedly occurred at Watkins Mill, but no documentary records of such are preserved in the firm's files.

According to a study of the legal culture of the era, "Emerging concepts of nineteenth-century tort law freed employers of responsibility for accidents involving their employees." Two brief notes from 1865 and 1874 indicate that, at Watkins Mill, fellow employees, rather than the employer, bore the costs of injured colleagues.[13]

Paternalism could be ambivalent when it came to balancing the workers' well-being against the bottom line. Controlling the workplace made it easier to exploit labor, as when eastern textile magnates used strategies to speed up as well as stretch out labor. At Watkins Mill, when new machinery came into place or the old was rearranged, the work processes often became more difficult or inconvenient for workers. Only two attempted remedies are now apparent: repositioning a trapdoor above the spinners to minimize handling of materials from the floor above, and cutting a new window in the exterior door to provide more daylight in the finishing room. Furthermore, the building was, by contemporaneous eastern standards, a firetrap. Perhaps through ignorance, Watkins did not use slow-burn construction designs for floors and walls, and he failed to provide fire escapes or any firefighting apparatus. (Fortunately no fire was ever reported during the factory's period of operation.) Such artifactual evidence gives clues that Watkins too was ambivalent in his paternalism.[14]

Compensation

During the years of Watkins Mill's operation, the average daily wage in woolen manufacturing rose nationally from a base index number of 100 in 1860 to 144 in 1880. This equaled figures for all industries and exceeded earnings for unskilled male workers, but fell behind cotton and gingham textiles. Economists and historians studying trends and labor practices at Watkins Mill have noted that the employees' pay

Figure 8.1: Fluctuations in the purchasing power of Watkins Mill wages between 1861 and 1898. Workers were paid $1–$3 per day. The graph shows the relative value of the $1 daily wage at different points in time, expressed as May 2000 hourly equivalents.

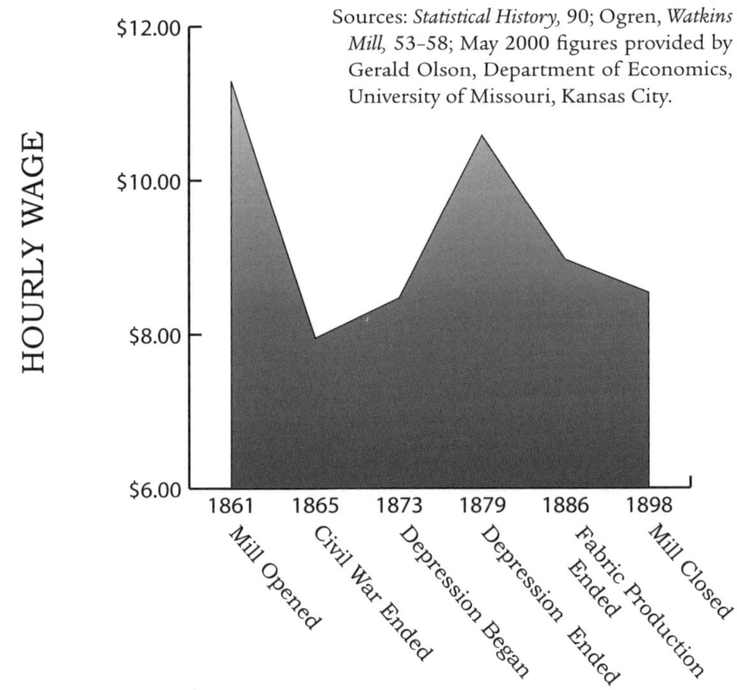

was comparable to their peers in midwestern mills. Highly skilled operatives like the dyer and the loom boss were usually the best paid, which may have contributed to their tendency to stay longer. Work in the mill generally paid better than jobs on the farm, and individuals who did both were paid at different rates. Waers was paid $2.50 per day for his mill tasks but $1.25 per day for hauling wagonloads of corn. In the days of dwindling factory operations, around 1890, Joe and Will Marley normally worked outside the mill at 77¢ and occasionally inside at $1.00 per day. In comparison, Watkins's weavers received 3¢–9¢ per yard and could complete 25 yards per loom in a day. Alternatively they could draw in two or three warps at 27¢ per warp.[15]

Although throughout the life of the mill, the workers were paid at substantially the same rates—between $1 and $3 per day—the value of those wages changed according to the prevailing economic conditions. Nationwide the cost of living declined during the nineteenth century, with price disparities varying regionally. As Cooper and Borden explain, "The value of labor has risen *faster* than the cost of goods, because the greater productivity of labor has made a unit of labor *more*

Three workers (from left): Will Marley, M. D. Goodwin, Joe Marley. According to the 1870 census, both Marleys were listed as mill hands. William was then age 16. Later in their careers, the duo worked both in the mill and on the Watkins farm. DNR, WMSHS.

valuable relative to the volume of goods and services that unit of labor can produce." Figure 8.1 shows fluctuations in the purchasing power of the wages paid at Watkins Mill.[16]

Workers might be paid in goods rather than in cash, another indicator that Watkins did not fully embrace the modernized economy. At Watkins's store—not a company store in the strictest sense—purchases of meats or woolen goods were recorded, as were expenses incurred by laborers at dry good merchants such as J. W. Asbury or Clark and Cummins in the surrounding communities. Ogren surmises, "Dry goods became the worker's earnings and the store owners received their payment as bolts of cloth or pounds of yarn." For example, Frances Vermillion, 19-year-old daughter of a local carpenter, appears in both the weaver's book for September 1865–January 1866 and the general ledger for September 1865–July 1866. In the former, she had five entries for each month's wages. In the latter, her purchases and withdrawals included seven grocery items (of finely milled flour), four pieces of cloth, eight cash withdrawals, and a purchase at a local shoemaker. A comparison of the prices Watkins charged her for 4.5 yards of flannel and 2.75 yards of cassimere with the rates he paid weavers indicates she would have to work eleven days to cover her cloth purchases. (Watkins gave no employee discounts, nor does his ledger credit her with any accumulated interest for her earnings.) Such arrangements reflect the persistence of the barter economy; they also enabled Watkins to survive when cash was unavailable.[17]

A small leather-covered book known as the "Sallie Turner Employee Expense Book" (dated 1874) is particularly prized among the Watkins Mill site records because so few documents remain that give the perspective of the employee. Similar to daybooks kept by Waltus or

John Watkins, this 4-inch by 3.5-inch book kept track of the weaver's credits and debits. Although, as noted, weavers were generally paid on a piecework basis, Turner recorded her wages as timework. Other mill operatives probably also kept their own copies of their personal accounts and thereby checked the accuracy of the ledgers kept by William Waers, the company's bookkeeper. Every two months or so, the accounts would be balanced and Sallie paid in cash due her. This personal ledger provides further evidence of the transition from the barter system of preindustrial societies to the cash nexus of an industrialized labor marketplace.[18]

Mill Management

The management of the mill reflected not only the Watkinses' paternalism, but also the transition to the new industrial society. The Watkinses, as heralds of the industrial value system, sought to alter the work habits, expectations, and aspirations of their employees. The late Herbert Gutman, one of America's foremost students of labor history, has explained, "In the half-century after 1843 industrial development radically transformed the earlier American social structure; and during the Middle Period (an era not framed around the coming and aftermath of the Civil War) a profound tension existed between the older American preindustrial social structure and modernizing institutions that accompanied the development of industrial capitalism."[19]

In setting up a manager–operative hierarchy in the mill that paralleled the patriarchal family structure, Watkins eased the transition from home to work and from farm to factory. In their paternal role the proprietors, especially Waltus, frequented the workplace, were acquainted with all tasks assigned, and knew their workers as individuals.

A set of work rules dated 1 May 1871 has been found among the firm's papers. This handwritten document, which might have been posted or circulated among the mill operatives to ensure their compliance, illustrates how managers in industrial capitalism sought to create new human relationships. Watkins's rules would provide "the government of our Mills." The preface pointed out that the employer could profit and the employed would prosper "in proportion" as labor "is systematically and energetically applied"; the rules themselves made clear that it was the manager who would dictate the system, while the operatives supplied the energy. The document outlines general principles; no particular behaviors are sanctioned, nor are explicit penalties prescribed for noncompliance.[20]

The first rule stated that timework rather than piecework was to be the true measure of one's productivity. A supervisor, not the craftsmen, would determine how "the labor employed is profitably applied." The second rule reiterated that the superintendent would assign each employee "to perform_____ hours labor per day." The third regulation was most crucial: "At times when there is a slackness or suspension of

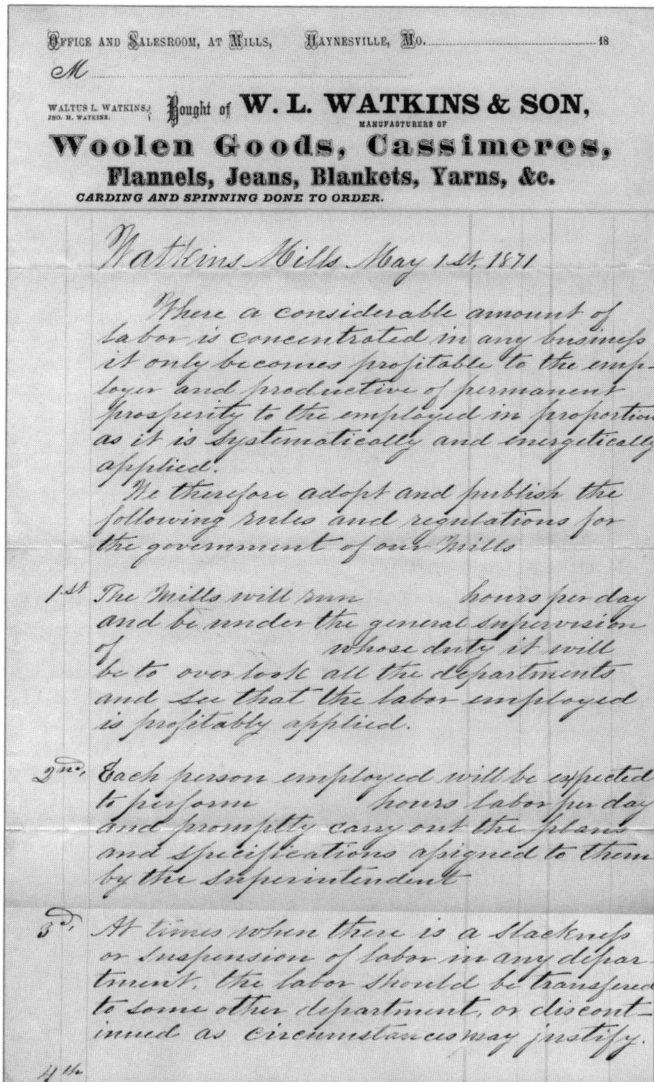

Office and Salesroom, at Mills, Haynesville, Mo. 18

M ....

Waltus L. Watkins, Jno. H. Watkins. Bought of **W. L. WATKINS & SON,**
MANUFACTURERS OF
Woolen Goods, Cassimeres,
Flannels, Jeans, Blankets, Yarns, &c.
CARDING AND SPINNING DONE TO ORDER.

Watkins Mills May 1st 1871

Where a considerable amount of labor is concentrated in any business it only becomes profitable to the employer and productive of permanent prosperity to the employed in proportion as it is systematically and energetically applied.

We therefore adopt and publish the following rules and regulations for the government of our Mills

1st The Mills will run hours per day and be under the general supervision of whose duty it will be to overlook all the departments and see that the labor employed is profitably applied.

2nd Each person employed will be expected to perform hours labor per day and promptly carry out the plans and specifications assigned to them by the Superintendent

3d At times when there is a slackness or suspension of labor in any department. the labor should be transferred to some other department, or discontinued as circumstances may justify.

4th

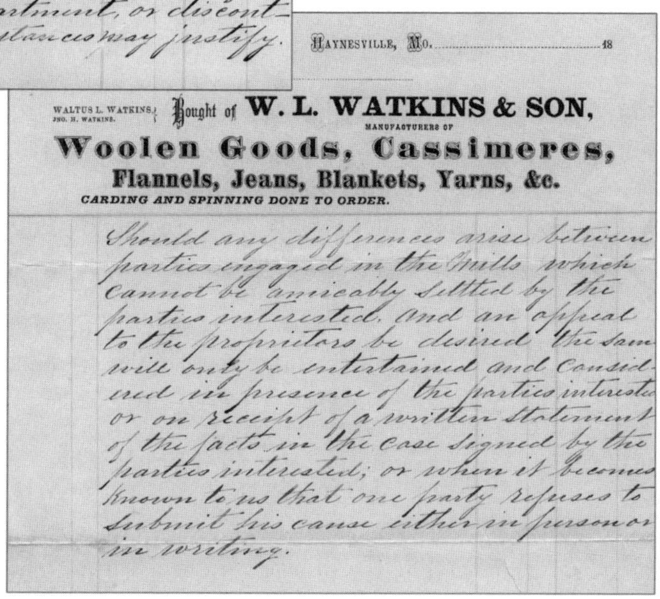

Haynesville, Mo. 18

Waltus L. Watkins, Jno. H. Watkins. Bought of **W. L. WATKINS & SON,**
MANUFACTURERS OF
Woolen Goods, Cassimeres,
Flannels, Jeans, Blankets, Yarns, &c.
CARDING AND SPINNING DONE TO ORDER.

Should any differences arise between parties engaged in the Mills which cannot be amicably settled by the parties interested. And an appeal to the proprietors be desired the same will only be entertained and considered in presence of the parties interested or on receipt of a written statement of the facts in the case signed by the parties interested; or when it becomes known to us that one party refuses to submit his cause either in person or in writing.

Rules and regulations governing Watkins Mills, 1 May 1871. DNR, WMSHS.

labor in any department the labor should be transferred to some other department, or discontinued as circumstances justify." This assumed that there were definable tasks—housed in departments—within the factory and that the production process was interdependent, yet not always continuous nor constant. Overall coordination was the prerogative of the manager, who had the option of *not* assigning work. As the mill was not a fulltime operation, Watkins needed to maintain flexibility in his labor resources. Working through his superintendent and other managers, he shifted personnel frequently among a number of posts within the factory and, as previously noted, moved workers from industrial operations to the farmstead. The third rule, as well as blank spaces in the rules document, accorded him this latitude.

The last rule provided for the proprietors to act as mediators in "any differences [that] arise between parties engaged in the Mills." This was not a grievance procedure between labor and management; rather, Watkins as proprietor would act as judge.

Unlike large-scale operations in the East, which operated according to printed and posted schedules, Watkins's daily regime apparently was not formalized in print. Thus it cannot be ascertained how much the factory operated according to a clock. Both the mill and the Big House had Seth Thomas clocks, enabling Waltus from his salesroom-office at the west end of the factory's ground floor and Mary Ann from her kitchen to time their daily activities. Watkins, and most probably his supervisors, possessed pocket watches. An 1867 announcement declared, "Hands are required to work from the blowing of the whistle in the morning to the whistle in the evening." It can be hypothesized that the entire labor force broke for lunch at the same time, idling the machinery.[21]

Control of the whistle was evidently a source of discord in the factory. Engineer L. H. Jones, who ruled the engine room in the 1870s, took offense at someone else's blowing of the whistle, apparently interpreting the action as both an accusation and a usurpation of his power. In an angry note, probably delivered to Watkins or posted, he wrote, "Notice, Notice…as I am bard [barred] out never can see what time by the clock I do the best that I can on Starting on time of a morning." He warned all to stay away from the engine room: "If that whistle is sounded eny more by eny body matters not who it is I will never touch a throttle again.… I mean what I say." The management reaction to the situation is not known, but as Jones is never heard from again and is not included in later records, it seems safe to say he left the mill not long after the controversy.[22]

Social Control and the Mill Community

From his earliest days at Bethany, Watkins had been shaping his social as well as his physical surroundings, envisioning a harmonious whole in which farm, factory, church, and school were blended together under Squire Watkins's paternalistic management. In providing housing to some of the mill employees, he literally created a community in which the members' relationships to each other were shaped by their employment at his factory. After an eleven-hour day in the factory, many operatives spent the remainder of their waking hours with fellow workers. The Jacksons, Walkers, Waerses, and other families shared communal experiences, moral precepts, and work discipline. They attended the Mount Vernon Missionary Baptist Church together, and their offspring intermarried.[23] These close connections, centered in tradition, could prove reinforcing as well, with workers' family members and friends perhaps providing some leverage against a supervisor's demands at work.[24]

The homes provided to the operatives as part of their compensation were arranged along a rise between the mill ponds and the state road. Collectively these buildings (nine to twelve in all) became known as "Smokey Row," as travelers passing by the hamlet glimpsed the telltale signs of the houses' chimneys set against the factory stack.[25] Local lore refers to these as "weavers' cottages" and indicates that those south of the Lane's gate were log cabins, while those on the north side were frame structures (possibly of more recent construction). Most houses consisted of one square room, about 15 to 16 feet on a side, with an additional 9-foot lean-to across the back wall; archaeological research suggests that the buildings on the south side may have included a barracklike structure. Further investigation may lead to revised estimates of how many people lived near the mill or in individual dwellings.[26]

Future searches might also discover more of other workers' homes scattered about the Watkinses' holdings. One such house is known to have been located on the Bethany Road near the current Visitor Center parking lot; it was at one time inhabited by Mary and Esseks Reeves, a Black couple. The Cavender Cabin was also on the Bethany Road, southeast of the modern railway. This was the home of John Jackson and Inez Waers Jackson in the 1870s and later of the family of Bill Cavender, who operated one of the Watkinses' sawmills next to the iron bridge over Williams Creek.

One unanswered question is whether slaves or free Blacks ever lived in the log cabins. Unverified tales claim that slaves inhabited some of the structures south of the Lane's intersection with the Bethany Road. No corroborating evidence has been found to support the assertion in the centennial history of Clay County (1922) that "negroes['] ... cabins were thick on the farm."[27]

Two families can serve as examples of the interrelationships and varied roles of the people connected to Watkins Mill. William Waers, the dyemaster and bookkeeper, lived with his eleven children in a house south of the factory, set apart from the other cabins. According

Gil Jackson (seated center) personified the spread of the industrial revolution. He learned skills as a woolen finisher in Britain, emigrated to the United States, and went to work at Watkins Mill where he became a long-time employee. In addition, at least two of his offspring worked at Watkins Mill. DNR, WMSHS.

to archaeologists, his home was larger than the usual accommodations for factory workers. He also owned land of his own: 7 acres of woodland (valued at $600) incorporated with a 40-acre farm operated by one of his sons due east of the Mount Vernon Church, and an additional 74 acres purchased with a loan from the church's trust fund. Many of Waers's children worked in the woolen factory. His oldest son, John, married Martha "Susan" Walker, daughter of the mill engineer; his daughter Inez, a weaver, married John Jackson, the fuller's son.[28]

The Jackson family's connection to Watkins Mill is another illustration of the middle ranks of the community there. Patriarch Gil Jackson was born in 1828 in England. There he trained as a fuller, married Amelia Hudson (1849), and had two children, Isabella (born 1855) and John (1858), before immigrating to the United States in 1858. Another son, William, was born in October 1860 following the family's move to Missouri. By the time of the 1870 census, Jackson's position at Watkins Mill had enabled him to accumulate realty holdings of $2,000 and personal property worth $200. According to the 1877 atlas, Jackson owned 40 acres south of Greenville. Employment records for 1872 show that Belle had joined her father in the factory, where she worked as a weaver. The community lived on in the person of William Ernest Jackson, a grandson of John and Inez Waers Jackson, who resided adjacent to the historic site until his death in 2002.

The Watkinses were part of this community and yet removed from it. Waltus Watkins commanded wealth, status, and political power in the hamlet. He dominated the significant social institutions. His children did not marry into employees' families (or other nearby families, for that matter). He shaped and controlled the others' lives, using the school, church, and other social entities to reinforce codes of behavior and worldviews he found acceptable in his workforce. Like the weavers at his looms, Waltus Watkins supervised the fabric of the community as he wove the threads of family life and of work, of the factory and the farm, of interlocked institutions.[29]

Labor Relations

The blowing of the factory whistle signaled the start of another workday for the factory operatives. Entering the factory meant accepting, in the words of Jack Larkin, "a world of precisely structured hours, close confinement and the noise of machinery.... It took a month or so to become reasonably comfortable with the machinery. Success depended on vigilance, quickness and deft manipulation. Some never acquired it and had to give up the work." The managers inculcated discipline. Whereas farm implements and other tools could be controlled by whoever wielded them, the textile machine was automatic. Thomas Bender observes, "The machine is thus more than labor-saving; it becomes life-shaping. The physical aspects of labor become less important than the psychic consequences of the workers' passive relationship to a mechanical production system, particularly under conditions of extreme division of labor." Ralph Waldo Emerson commented that machines "have some questionable properties. They are reagents. Machinery is aggressive. The weaver becomes a web, the machinist a machine.... All tools are in one sense edge-tools, and dangerous."[30]

Within the factory, Watkins and his employees entered "contested terrain." Analyst Merritt Roe Smith has said of such Industrial Revolution sites, "People found themselves working for employers who owned the tools of production and who insisted on dictating how, when, and where the work should be executed.... The question of who should control production became a fundamental point of contention in the 19th-century industrial world." The outcome did not necessarily involve direct confrontation; Smith further observes that in "the day-to-day interactions that took place between workers and managers...we find complex, subtle relationships that blend deference with defiance and suggest action among equals as well as an emerging sense of class differentiation."[31]

As Missouri industrialized, strife between management and labor arose. In the summer of 1877, a strike of eastern railroad workers, whose salaries had been cut due to the panic of 1873, spread to Missouri. In the end of July the economy of St. Louis was brought "to a standstill, among other things closing at least sixty factories. The threat of violence was everywhere."[32] Sedalia, Hannibal, and St. Joseph also suffered. Labor groups in St. Louis sought an eight-hour day and curtailment of the employment of children under the age of 14. However, these were apparently not contentious issues at Watkins Mill; evidence suggests the mill was ruled by a sense of community rather than adversity.[33]

Just as Watkins had located his industrial enterprise within his agricultural endeavors—the middle landscape—so he sought middle ground in the give and take of sparring over wages, hours, and working conditions.[34] Watkins did not view the situation as a confrontation between traditional habits of workers and new demands of the Industrial Revolution. He, in fact, had no superior control of the variables in bargaining with workers. A number of clues indicate that he consciously avoided confrontations with his employees (although, as the incident with engineer Jones and the whistle demonstrates, it cannot be assumed that the workers were docile). First, Watkins's labor recruitment strategies emphasized factors—especially his rural setting, bias toward generalists, and strict moral code—most likely to turn away potential adversaries. Second, his personal bent toward experimentation in a number of enterprises meant he chose not to be linked with other woolen magnates and become competitive for profits. Third, in his penchant for not purchasing updated equipment, he continued his dependency on skilled workers; Laurence Gross hypothesizes that at Watkins Mill, "a large part of the operation" was removed "from direct control of management." Finally, the employment of families maintained order in the factory as well as at home.[35]

The absence of direct evidence of major labor problems does not mean there was no unrest. Nor were the workers powerless to affect their working conditions. In the words of Merritt Roe Smith, "What we see time and again are everyday people developing a variety of seemingly simple and ad hoc methods of mollifying the power of their employers." Pacing and absenteeism were resistance tactics that employees often invoked, though it is not known how often these came into play at Watkins Mill. Perhaps the greatest measures of dissatisfaction at Watkins Mill are letters spurning job offers and the turnover in labor from one year to the next. Such turnover was common in the woolen industry, reaching levels as high as 45 percent per year. Rates were lower in rural areas, perhaps because, as a U.S. Tariff Board survey of the industry for 1870–1910 reported, "many of the (American) woolen and worsted mills are located in small villages and towns remote from large cities or large centers in which other industries are carried on. In such localities there is more steadiness and permanence of employment in the mills than in larger centers." One study of New England mills pegs the average tenure at a mill position as 1.1 to 1.4 years.[36]

Because labor records at Watkins Mill lack continuity, it is impossible to make definitive statements about turnover rates there. One set of records lists twenty-six different workers for 1866 (nine months), forty-four for 1867 (ten months), and forty-three for 1870. Ten of these individuals appear on both the 1867 and the 1870 lists. If this sample is representative, it would show Watkins retaining approximately 23 percent of his workers for at least three years. As noted earlier, a number of employees remained in the Watkinses' employ for fifteen or twenty years.[37]

One area for potential strife—the threat to workers' jobs posed by upgraded machinery—was mitigated by circumstances at Watkins Mill. First, technological innovation did not necessarily lead to job displacement, nor to less skilled workers. As machines required skilled technicians to run and maintain them, some workers, especially those whose skills included diagnosis of machine malfunction, enhanced their status. Gross's study of wool carders notes, "The carder's abilities increased despite machines devised to diminish skill. The roles of his associates will be seen to have changed little, while the machine-maker's skill was more successfully reduced." Second, when Watkins had a choice, his preference was not to install more efficient machinery but to retain skilled operatives. This strategy may have contributed to more harmonious labor relations, but it also made the enterprise less competitive (additional evidence that Watkins was not a strictly profit-centered industrialist, but an opportunist).[38]

The truth about the labor situation at Watkins Mill must be pieced together from artifactual evidence and documentary sources. Doing so leads to certain conclusions: The rural location and Watkins's limited commitment to milling as a part-time, seasonal, and civic endeavor resulted in a melding of agrarian and industrial life. The owners' preference for hiring family-related operatives reinforced traditional mechanisms of discipline and control shared between home and work. Impersonal contracts did not play a major part in structuring society in northeast Clay County. Shortages of skilled labor, compounded by reliance on outmoded machinery, locked Waltus and John Watkins into a system that was conducive to workers' autonomy. The provision of multiple institutions—mill, school, church, and cabins—at Bethany can be viewed as an effort by Waltus to create a self-contained, if not self-sufficient, community. Finally, the relationships among Waltus and his employees were permeated by a paternalism, common to many nineteenth-century textile mills, that both aided the smooth functioning of the mill and undermined its ability to compete in the widening marketplace.

NOTES FOR CHAPTER 8

[1]Brown, *Modernization,* 3–22; Crockett, *Woolen Industry,* 95.

[2]Scranton, "Varieties of Paternalism," 236–39; Wingerd, "Rethinking Paternalism," 872.

[3]Watkins did not employ members of his own extended family in the factory, except for four sons who worked there briefly on their way to managerial positions, nor did he let his children marry any workers.

[4]Prude, *Coming of Industrial Order,* xii; Thelen, *Paths of Resistance,* 3.

[5]Waltus Watkins to John Byron, 17 February 1879, John Watkins and Brothers to F. Ray, 15 April 1884, WMSHS.

[6]James Ash to John H. Watkins and Brothers, 29 January 1884, WMSHS; John Watkins to F. Ray, 5 April 1884; employee day book, WMSHS. Unfortunately it is impossible to know whether the applicants' letters are reliable measures of the workers' actual skills and conduct.

[7]J. D. Hall, R. Korstad, and J. Leloudis, "Cotton Mill People," 246.

[8]Ogren, *Watkins Mill,* 59–63.

[9]Population Schedules, Seventh–Tenth U.S. Censuses, 1850–1890; "Employees," interpretive manual, WMSHS; Licht, *Industrializing America,* 64–65.

[10]Gordon and Malone, *Texture of Industry,* 384–85.

[11]Rucker, Interview.

[12]Ogren, *Watkins Mill,* 63; "Watkins Woolen Factory Personnel: 1867 Night Carding," WMA Collection, WMSHS.

[13]K. Hall, *Magic Mirror;* 135; daybook, 1873–74, WMSHS.

[14]A threatening grass fire on the Lane in the late 1950s did have to be extinguished by Lawson volunteers and local history buffs. Gross, "Importance of Research," 25; Oberholtz, Interviews by Potts, 23 June 1987, 30 June 1987, and 29 February 1988.

[15]*Statistical History,* 90; J. C. Cooper and K. Borden, "Interpretation of Wages"; Ogren, *Watkins Mill,* 53–63; Crockett, *Woolen Industry,* 63–64. There were too few male weavers at Watkins Mill for a valid comparison, but records for the industry as a whole show that they were generally paid double what female weavers earned. Cole, *American Wool Manufacture,* 1:240.

[16]Ogren, *Watkins Mill,* 53–63; J. C. Cooper and K. Borden, "Interpretation of Wages," 19.

[17]Ogren, *Watkins Mill,* 16; Lentz and Wilson, "'Mill Girls."

[18]Sallie Turner employee expense book, 1874, WMSHS.

[19]Gutman, *Work, Culture and Society,* 13.

[20]Business correspondence, 1870–73, WMSHS.

[21]Letters or autobiographical statements concerning the Watkins family members, typescript, Research Files, WMSHS.

[22]Letters or autobiographical statements concerning the Watkins family members, typescript, Research Files, WMSHS.

[23]The sense of community shared among mill workers and neighboring farmers, as well as Watkins's paternalism, is seen clearly in the minute book (1865–87 vol.) of the Mount Vernon Missionary Baptist Church (WMSHS). See also Potts, "Waves of Revivalism."

[24]J. D. Hall et al., *Like A Family,* 114–80.

[25]Bray and Yelton, *Archaeological Investigations.* A fictional account can be found in Entrikin, *Smokey Row.*

[26]*Archaeological Investigations: Waers Cabin,* 117. Documentary records can also aid in estimating the numbers of residents. Ledgers kept at the mill store for 1865–66 indicate that the Toler family charged extraordinarily large amounts of foodstuffs against the wages of four family members who labored in the factory. Perhaps they accepted boarders.

[27]"No negroes were employed in the mill, though their cabins were thick on the farm, and they helped there and in the home." DAR, *Centennial Souvenir.* The statement that Watkins did not employ Blacks in the mill, although likely true, has also not been verified.

[28]*Archaeological Investigations: Waers Cabin.*

[29]Perhaps most susceptible to Watkins' formal control were those workers who boarded with the Watkins household. In the fall of 1862, fifteen workmen, among them the elderly store clerk and a jack spinner or two, took their meals in the Big House.

[30]Larkin, *Reshaping of Everyday Life,* 56; Bender, *Toward an Urban Vision,* 61, 58; Emerson, "Works and Days," 164.

[31]M. R. Smith and G. Clancey, *Major Problems,* 183–84.

[32]Christensen and Kremer, *History of Missouri,* 12–13.

[33]William Parrish notes that the state legislature in 1867 declared eight hours to be the standard unit of a day's work if agreed to by both labor and management. Loopholes and exceptions prevailed. Parrish, *Missouri Under Radical Rule,* 224–25.

[34]Richard White asserts, "Perhaps the central and defining aspect of the middle ground was the willingness of those who created it to justify their own actions in terms of what they perceived to be their partners' cultural premises. Those operations in the middle ground acted for interests derived from their culture, but they had to convince people of another culture that some mutual action was fair and legitimate." White, *Middle Ground,* 52.

[35]Gross, "Importance of Research," 20–23; Gross, "Watkins Mill," 71–75.

[36]M. R. Smith "Industry, Technology"; M. R. Smith and G. Clancey, *Major Problems,* 182–89; Cole, *American Wool Manufacture,* 118–19; Prude, *Coming of Industrial Order,* 89, 145.

[37]"Watkins Woolen Factory Personnel: 1866 Workers, 1867 Workers, 1867 Night Carding," "Personnel for 1870," WMA Collection, WMSHS.

[38]Gross, "Wool Carding," 805.

9

THE LAST TIME WE'LL TURN THE WHEELS

Watkins Mill, as a woolen factory, did not long survive the death of its founder. Even before Waltus Watkins succumbed to a lengthy illness and "softening of the brain" in 1884, the factory had started a downward slide.[1] Its decline and eventual demise can be attributed to a number of factors. Some of these could perhaps have been avoided if the owner-managers had taken a different approach to the business; others resulted from circumstances that were either inherent to their situation or widespread throughout the industry and thus largely beyond their control.

DECLINE IN THE WOOL INDUSTRY

The peak of midwestern woolen manufacturing came in 1870, when the 881 establishments then operating in the eight-state region comprised 36 percent of the nation's woolen factories. Before long, however, a combination of local factors and more widespread changes in the markets led to an irreversible decline. In its thirty-year existence, the midwestern woolen industry had provided a transition from the cottage or handicraft industry to the modern, centrally located, factory system of mass production. Now the residentiary firm with its regional market was displaced by larger-scale operations integrated into a national market. Cloth production and finished goods were dominated by large woolen mills in the hands of big business. The railroad lines that brought the mill its supplies also brought less expensive cloth from big eastern mills. In addition, consumers' clothing demands and purchasing habits changed. The advent of central heat in American homes affected clothing tastes; in particular, worsteds and cottons proved more attractive to consumers in search of lighter-weight clothing. Department stores in cities, as well as mail order catalogs throughout rural America, offered ready-to-wear garments. Consumers were lured by fashion, price, and credit to purchase clothing rather than make their own. Between 1873 and 1898, wholesale prices of woolens fell faster than other commodities. By 1900, only 183 woolen factories remained in operation in the Midwest.[2]

Figure 9.1: Trends in the number of woolen mills operating in Missouri during the final years of Watkins Mill. The figures for Fabric Mills include those with cloth-producing capability, not just carding, spinning, or fulling services. Total Mills includes all woolen mills, from carding-only facilities to full-production factories. The sharp decline between 1885 and 1887 coincides with John Watkins's decision to stop fabric production at Watkins Mill.

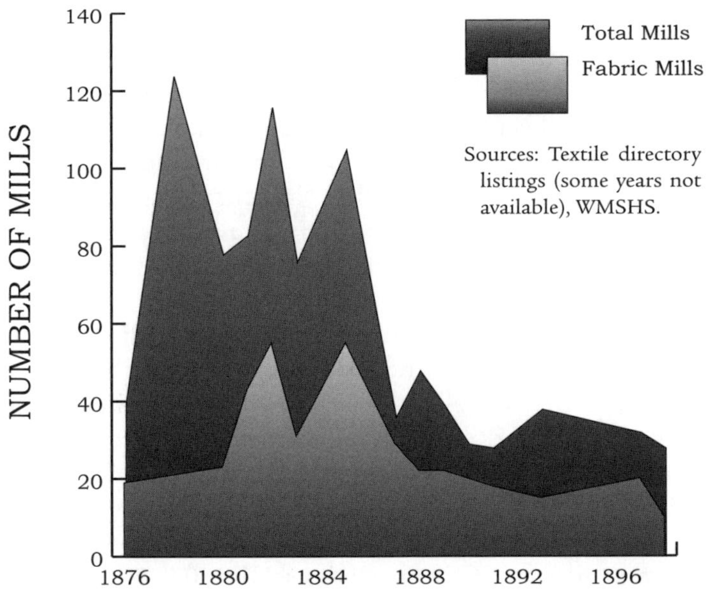

In Missouri, sheep production fell and was reoriented to the meat market rather than the fleece trade. By the end of the 1890s, only twenty mills still operated, and only half of those produced cloth. "Faced with declining local demand for piece goods and yarn," analyst Norman Crockett has observed, "most of these older woolen men simply closed the doors of their factories, accepting what appeared to be inevitable"—a generalization that fits the actions of John Watkins, Waltus's principal heir.[3]

DEATH OF WALTUS WATKINS

Waltus Watkins died on 24 January 1884, after an undetermined period of failing health. Following a trip to Kentucky in 1881, the family saw his condition decline; as early as January 1882, they were reporting that he looked "feeble." Their correspondence refers to his having "been afflicted with paralysis for several years" and undergoing "severe suffering for such a length of time."[4] Carrie wrote dejectedly to her sister that "this hill is as lonesome as a wilderness I feel like I was isolated from the whole world.... [At church] our dear fathers place was vacant it was indeed a sad scene and our pastor gave us a sad sermon, I fear 'twill be many days before the church has such a pillar of support as father." Carrie's feelings of isolation were no doubt compounded by deaths of two other church members who were also neighbors, members of the Waers and Jarrett families, at about the same time.[5]

A number of transactions made during Waltus's final years provided for the senior Watkinses' support and also ensured the smooth transfer of assets to their heirs. On 25 September 1882, a deed conveyed all Clay County landholdings (a total of 2,228.7 acres) from Waltus and Mary Ann to their sons John, Joe, and Judson. "A certain contract" noted in the deed was evidently executed simultaneously, although it has not been found. A little more than a month later, lands in Ray County (1,442 acres) were likewise transferred from one generation to the next. Both deeds specified "in consideration of the sum of one dollar." Reserved from the transfer was a lifetime residence at the Big House for the parents; indeed, Mary Ann would live at the estate until her death on 24 September 1896.[6]

Ledgers dating from September 1882 indicate that the family in effect created a living trust. The sons kept track of the costs of supporting their parents, as well as of cash payments on their behalf from the newly designated firm of John H. Watkins and Brothers. From 25 September 1882 until 31 December 1883, support for Waltus and Mary Ann totaled $2,537.57 (including such items as $37.50 for a new easy chair for Waltus and $10.00 for vases and flowerpots for Mary Ann). In the same period, Waltus advanced $407.55 in cash and services. As researcher Booker Rucker has deduced, the existence of this ledger explains why a will was never probated and implies the transfer of the woolen company. These conclusions grow out of the 1 January 1884 entry that Waltus's apparent indebtedness of $2,130.22 was more than balanced by a notation that the boys owed $2,527.75 "interest on contract" to their father. The missing "certain contract" of 1882 covered the woolen business, the value of which was large enough to generate over $2,500 interest in fifteen months. Thus at the end of 1883, Waltus was really a creditor of his three sons; they issued a new note ($397.73) to that effect.[7]

Entries for early 1884 itemize costs for Waltus's final days: medical care by Dr. James of Lawson; burial clothes ordered from general merchants Asbury and Brothers; the undertaker with his hearse for the funeral and burial at the family plot. A final notation, dated 5 February 1884, reveals the existence of Waltus's last asset, a $25,000 note credited to his estate by John Watkins and Brothers; with interest, Mary Ann and the couple's eight children would each receive $2,818.49 from the firm.

Booker H. Rucker discovered a note (originally found in Mary Ann's pocketbook, presumably after her death) listing the liabilities and assets assumed by the three Watkins sons when they took over their father's affairs. Between taxes and debts to a dozen creditors,

Waltus Watkins had laid the groundwork for the image that has been perpetuated when he contributed self-appraisals for articles in the *Liberty Tribune* (1870) and the *United States Biographical Dictionary and Portrait Gallery of Eminent and Self-Made Men, Missouri Volume* (1878). In the dictionary Waltus emphasized for posterity his family's Welsh and Huguenot heritage, his extensive training with machinery as a young man, and his triumphs over adversities as a farmer and businessman in Missouri. A lengthy paragraph outlined how well his offspring—all having "received a good education"—had married or commenced careers. Two of the Watkins girls had married nephews of Senator David Rice Atchison; their brothers were either promising stockbreeders, farmers, factory managers, or students at William Jewell College. Particularly highlighted as Waltus's successor was John, "a thorough and active businessman" in partnership with his father.[a]

John Watkins. DNR, WMSHS.

[a]"Old Men of Clay County"; *U.S. Biographical Dictionary.*

Waltus owed a total of $56,000. In 1882, he had five major creditors: W. W. Smith (Judson's future father-in-law), $13,200; son-in-law M. D. Scruggs, $8,600; Clay County, $7,500; the famous Alexander Doniphan, now a Richmond banker, $6,000; and former employee John Donahue, $5,000. Some of these debts might be traceable to the failure of Watkins's investment in Ray County coal mining; others represented loans and mortgages by which he had secured capital for his various ventures.

The itemized assets fell into three categories: 3,600 acres of land at $15 per acre or $54,000; $18,900 in livestock and corn; and $7,100 in factory products. The bottom line was a positive balance of $24,000 after liabilities were subtracted from assets. The list leaves unanswered the questions of what value was placed on the factory itself and why the land was so undervalued—neighboring plots were then valued at

$21 or more per acre, and within two years, the sons sold some of their acquisitions at $250 per acre.[8] In any event, the fact that Watkins's agricultural pursuits comprised 73 percent of his listed resources (not including real estate) while manufacturing contributed 27 percent may be a reflection of his priorities.

Posthumous appraisals of Waltus Watkins have been found in two printed eulogies and in obituaries that appeared in church minutes and local newspapers. One of the eulogists, perhaps pastor J. J. Felts, emphasized Watkins's secular contributions, noting, "He had been a man of great energy, and his life had been a busy one.... No better citizen ever lived or died in Clay County." The other was offered by his former minister, W. C. Barrett, who officiated at Watkins's funeral; it focused on his religious life, observing also that "he was successful in business and liberal to a fault." These sources formed the basis for entries in later county histories, which placed more emphasis on Watkins's character and on the variety of his activities. His role as a local booster was particularly noted. One writer emphasized that "the many interests in whose advancements and promotion he assisted will long keep his memory green in the hearts of friends and neighbors." As noted in chapter 3, Watkins purposely did not charge high prices in his land sales in order to encourage settlement in the community. He was prominent in all "the best interests of the county."[9]

Demise of Watkins Mill

In 1886, the cloth shear in the Watkins finishing room broke irreparably. This relatively simple machine had failed twice before and been sent back to the manufacturer in Housick Falls, New York, for repair. Now the manufacturer said the shear was beyond repair and recommended that John Watkins purchase another model. The woolen trade in Missouri having already fallen precipitously, John declined. As a result, textile production came to an abrupt halt; yarn production, which did not require a shear, continued until 1892. After that the steam engine would be started up sporadically to handle a custom job in carding, spinning, or fulling some local wool, but even those jobs ceased altogether in 1898. The engine continued to power the gristmill until 1904; when it broke down, the factory finally closed for good. Inventories of yarn, blankets, and other goods continued to be available from the mill's store. (Some leftovers were sold as late as 1956!)[10]

Though the folklore of Watkins Mill attributes the mill's demise to the failure of the shear, the truth is not so simple. Failure to compete amid declining markets, inability to recruit effectively for skilled workers, inefficiency in the layout of the factory, reliance on all-too-distant services for some machinery maintenance, loyalty to outmoded equipment, diversions in agricultural pursuits—all contributed to the mill's difficulties. Managerial shortcomings reflected the Watkinses' lack of

Mill store, northwest corner, first floor. Photo by Jet Lowe, 1978 (78).

commitment to their industrial venture as a capitalist undertaking. Paul Ogren appraised the situation in this way: "The major problem of the mill was that this commitment to manufactures was strong in theory, yet limited in practice by the type of vision which the father and his sons pursued.... So the Watkinses were caught by much larger economic circumstances than the ones they thought they were dealing with."[11]

Marketing Decisions

The panic of 1873 set off five years of deflation and a severe depression throughout America, the longest in the nineteenth century. With cloth sales declining for almost two years, the Watkins woolen factory evidently shut down in 1874–75; between September 1874 and August 1875, only seventeen sales were recorded. During this time, some new equipment, such as a yarn-twister, was installed, and older machinery was rearranged within the building. When production resumed, it was much more slanted toward yarn, which required less labor and machinery and was therefore cheaper to manufacture. In the last half of 1876, Tootle and Hanna (Kansas City) placed a single order for 1,400 pounds of mixed yarns, which exceeded all yarn sales in the previous year. Between August 1876 and late 1879, yarn would gradually displace cloth as the leader of the Watkins Mill product line.

One product exemplifies the general decline in cloth prices over the lifetime of the Watkinses' firms. White flannel sold for 60¢ a yard in 1862. It rose to 90¢ by 1865, fell back to 65¢ in 1868, and dropped to 40¢ in 1870. By the end of the 1870s, which was racked by economic setbacks, the same cloth sold at 35¢ a yard. Correspondence with merchant-agents in the period 1874–85 indicates that even these lower prices were too high in comparison with eastern goods. Watkins agreed to reduce prices, but his sales volume continued to drop. Utilizing better economies of scale, manufacturing efficiencies, or cheaper raw wool, Watkins could cope in the early stages of this declining market. He could not compete in the later stage.[12]

As cloth production declined quantitatively, the Watkinses also cut back on varieties. A later drop in quality appears tied to their increasing reliance on marginally trained farmhands in textile production.

John Watkins, who could perhaps be described by the term "landed gentry," went on to become a most genial squire. Though he never liked to hunt, from 1909 until his death in 1931 he hosted the annual meetings of the Missouri Valley Fox Hunters Association. These paradoxical affairs looked to the past (by celebrating the baronial rights of the hunt) as well as to the future (by creating instant recreational communities at Bethany). Each autumn over 140 tents, some as large as five bedrooms, were raised. These were connected to a restaurant, movie theater, barber shop, and other amenities by means of a quarter-mile-long electrically lighted street, complete with telephone line,. A permanent kennel for 240 hounds was erected on a ridge south of the orchard. One observer opined that "a more perfect spot could not have been chosen than [Bethany] then was, thickly carpeted in bluegrass and generously shaded by towering virgin trees."

In addition to hunts, there were dog shows, horse shows, concerts, and other contests and exhibitions of the type now featured in county and state fairs. A platform and reviewing stand served as the center of the community, at the crossing of Fox Trot Boulevard and Hound Dog Avenue. According to newspaper accounts, one afternoon was set aside "for fox hunt guests to tour the old woolen factory they had driven past to reach the camp grounds. On such occasions, local citizens who had worked in their youth were called on to give carding and spinning demonstrations." The week-long event, reportedly the largest of its kind in the United States, attracted in excess of twenty thousand people, with as many as five thousand in temporary residence.[a]

Ready for the hunt. DNR, WMSHS.

[a]Roney, *From Entry Fee,* 80–81; "The Missouri Valley Fox Hunt," *Clay County Sesquicentennial,* 131–32.

The Watkinses' decisions to concentrate on yarn production and to deal with jobbers and wholesalers did not sit well with the small retailers they had served in the region. J. Porter in Plattsburg complained, "I do not exactly understand your mode of doing business. You claim that you are selling goods to me for less figures than you are to anyone else.... I do not understand how it is that jobbers are selling your yarns for less money than you are. I would like to hear from you on this subject."[13]

The production and marketing decisions that were made around 1881 were probably the work of John Watkins, as Waltus's old age and physical infirmities had by then led to his withdrawal. John had managed the mill for a decade. Along with Judson and Joe, he had taken some business courses, and all three had a lengthy yet narrow education in textiles, almost entirely in and around their father's woolen enterprise. They were not prepared to compete with ready-to-wear, store-bought, or catalog-hawked clothes produced by mammoth mills in the East and delivered throughout the nation by ever more efficient railroads.

As the cloth market had constricted in the 1870s, the yarn market fell off in the 1880s. The Watkinses upon occasion could not fill orders, nor could they cope with falling prices. In 1880, fine white yarn sold for 66¢ per pound and colored for 57¢, with yarn accounting for $15,000 in sales. In 1885, the prices stood at 52.5¢ and 45¢ respectively. The following year, production spurted to 16,200 pounds of yarn at $7,700, as prices briefly rose.[14]

In 1887 John Watkins halted production at the factory. Jobbers' orders ceased, inventories exceeded demand, and the carding and spinning machines did only occasional custom work. In 1889, Watkins Mill reentered the yarn market, recording $4,200 in sales, including an order from Marshall Field Company in Chicago. Only the jobber of Tootle and Hanna persisted, but by 1893—their last purchase—they sought only $180 worth of yarn. Ogren hypothesizes, "Perhaps [John] Watkins had decided on a bottom line and enforced it. Again the machines continued to run, but only for custom work.... The final phase of sales consisted entirely of custom work, remnant sales and perhaps minor production. But no longer would large orders motivate the Watkinses or their labor force."[15]

Outmoded Equipment

As previously noted, Waltus Watkins did not keep up with technological advances in the field of wool manufacturing and thus could not compete effectively with those who did. The best evidence of Watkins's industrial investment pattern in terms of machinery is what is *not* at Watkins Mill. The sets of cards were not improved by the so-called Apperly feeds available in the 1870s or the more expensive Bramwell feeds of the next decade. Gross believes rejection of the first "indicates a willingness to accept a high degree of labor intensiveness and of limitation in the quality of product." Spurning the second, "carries an implication of the availability of very inexpensive labor, but also a refusal to purchase the kind of equipment without which his position relative to other, more innovative, manufacturers could only slip backward." The first choice had been Waltus's; the second was probably John's. The Watkinses made similar choices with regard to other types of equipment—not purchasing an automatic spinning jack when those became available, for example.[16]

Ironically, the Watkinses' decisions about their machinery can be credited for that equipment being preserved for posterity—the best remaining example of an estimated 2,400 woolen mills of the era. Because no wool manufacturer would bid on the worn-out or obsolete machines when John sought to sell them, many remain today in their workplace, bearing witness to the nineteenth-century wool-making processes and industrial practices in which they were used.

Sons' Lack of Commitment

Crockett concludes that "the Watkins brothers lacked even their father's meager commitment to woolen manufacturing." Joe and Judson were partners in the venture for only about six years (1882–88). John, the eldest, worked alongside Waltus in managing the factory from at least as early as 1868. He held on awhile after Waltus's death, but it appears his interests lay elsewhere. Lacking his father's mechanical inclination and entrepreneurial drive, and given declining profits, John had little incentive for keeping the mill open. Before long he withdrew, closed operations, reverted to farming and stock breeding, and by the turn of the century had taken advantage of a new opportunity as a dealer of farm machinery.[17]

The Final Years

Watkins Mill had experienced a small and temporary revival during its years as John Watkins and Brothers (the firm's name from 1882–88). Some debts were paid by proceeds from land sales, and refurbishments were made in the mill. An 1881 handbill trumpeted that W. W. Watkins and Son was offering "great inducements in our Celebrated Woolen Goods" in exchange for raw wool. Two years later John Watkins and Brothers emphasized goods "for sale at the *Lowest Cash Prices,*" as well as "We make a specialty of Coil Rolls."[18] In 1888–89, the partnership was dissolved and division of the assets made, with John's portion including the factory and family home.

The figures that John Watkins submitted for the 1890 census indicate the moribund status of the mill. In a response drafted on company stationery, John estimated the value of the factory, including land, buildings, and machinery, at $30,000. The value of goods and services was $9,500, and his profit, at best, was less than $700. Table

Figure 9.2: Watkins Mill property value, products, and profits as reported in censuses of 1870–90

	1870	1880	1890
FACTORY PROPERTY			
Actual (2003 equivalent)	$65,000 ($941,816)	Not available	$30,000 ($611,777)
GOODS AND SERVICES			
Actual (2003 equivalent)	$45,000 ($652,026)	$20,000 ($379,724)	$9,500 ($193,730)
PROFIT			
Actual (2003 equivalent)	$6,900 ($99,977)	$3,220 ($46,656)	$688 ($14,030)

Sources: Industrial Schedules, Ninth and Tenth U.S. Censuses, 1870, 1880; John H. Watkins notes on Eleventh U.S. Census (1890), WMSHS; "Consumer Price Index (Estimate) 1800–2000," Federal Reserve Bank of Minneapolis, http://minneapolisfed.org/Research/data/us/calc/hist1800.cfm (accessed 24 February 2004).

9.1 shows these figures, together with their 1870 and 1880 counterparts (converted to 2003 dollars for purposes of comparison).

Even given the inexactitude of such comparisons (due to different reporting standards, discrepancies in accounting, and problems of figuring inflation and other economic variables), the figures do indicate a dramatic decline. In the same period of time, in spite of grasshopper plagues, droughts, deflation, and other hardships affecting farm profits, the proceeds of the Watkinses' agricultural endeavors increased by over 60 percent. John would have been foolish to further pursue wool manufacturing in the face of such numbers.[19]

Sixty-five years later, Charles Dagley (then 81 years old) reminisced about the final day of mill operations. The interviewer recorded, "On July 1, 1904, he began working for John Watkins.... The mill was not running, however a fairly large amount of wool and shelled corn were stored in the building." One morning in late summer or early fall, John directed Dagley and two other laborers to report to work. "On Friday the three (with John supervising) fired up the boiler—set the machinery in motion. Charley Cane was the engineer and Dagley and Marley ran the grist mill." After recess for the weekend, the men returned Monday. "John Watkins started operation and then left for Liberty. Joe Watkins more or less supervised the day's operation in his absence." This time the tasks were with the wool, Dagley was at the picker and Marley at the cards. "About halfway through the day, the engine broke

Watkins Mill as it appeared c. 1910. DNR, WMSHS.

down and the work had to stop. When John Watkins returned later in the evening, Joe said, 'John, the packing was blown out of the engine, what'll we do?' John replied, 'Well, I guess that's the last time we'll turn the wheels.' As far as Mr. Dagley knows, the mill was never run after that day." Dagley resumed his farm chores, which now included storing dismantled carriages in the factory.[20]

Decline of the Community

To a certain extent, the fate of the area surrounding the mill paralleled that of the Watkins enterprises, diminishing in some aspects and growing in others. In the 1890 census, the once packed and vibrant Watkins home now housed John (age 58) and Joe (41), their sister Carrie (45), and Joe's wife. (Judson is listed as living in Lawson with his wife and two daughters on a prosperous dairy farm.) The hamlet of workers' homes around the mill now lay mostly empty. As noted in chapter 4, the crossroads towns nearby had all but ceased to exist as population and commerce were drawn to depots along the railroads. The mood of many was perhaps expressed in 1881 by the Claytonville postmaster, who said: "Anyone who wants to be Claytonville postmaster can be, far as I'm concerned, just by holding up his hand."[21]

Lawson and Kearney—both on railroad lines—emerged as the commercial centers of the area. The Watkinses operated as many as seven brick kilns in Lawson between 1870 and the 1890s, and they made use of the St. Joseph Railroad railhead there for transporting raw materials and finished goods from the farm and factory. Starting in 1870, Watkins used a Lawson business address for his enterprises. Interestingly, it was he who in 1864 had sold the acreage that would eventually become the town of Lawson, perhaps indirectly contributing to the demise of communities closer to his homestead.

Thirty Years of Success

In his study of midwestern enterprises such as Watkins Mill, Norman Crockett has concluded that "the small, integrated woolen mill attempting to manufacture a complete line of coarse- and medium-quality cloth and serve a local market, symbolized a manufacturing unit of the past. Although some regions of the country still remained relatively isolated as late as 1900, the possibility of a profitable operation of these mills was vanishing.... [T]he tiny woolen mills of the Midwest fell victim to the very economic progress they had helped to create." Laurence Gross, who investigated the physical plant, equipment, and techniques at Watkins Mill, offers a different perspective: "Operating a great distance from the centers of the machinery, manufacture and labor he required, producing a wide variety of goods on a wide variety of machines, the Watkins Mill experience appears more remarkable for the degree of success achieved rather than for its ultimate failure." Certainly the two generations of Watkinses who managed the mill, when measured only by capitalistic criteria, failed in the long run. Yet measured on other scales, their enterprise was a success. Watkins Mill had a longer than average lifetime, largely due to Waltus Watkins's entrepreneurial drive, willingness to adapt to changing circumstances, and skill in maximizing his opportunities. For over thirty years, the factory provided goods, services, commerce, and employment to a large area in western Missouri. The proprietor of the mill also helped found a community and put his personal stamp on it and its members. Waltus Watkins and his "factory on the farm" represent the social and economic growth, transition, and values of nineteenth-century midwestern America. The factory shut down, but its legacy continues.[22]

Notes for Chapter 9

[1]Waltus Watkins death certificate, Clay County Archives.

[2]Crockett, *Woolen Industry,* 94ff; Gross, *Course of Industrial Decline.*

[3]The last woolen mill in Missouri, located in the town of California, burned in the late 1960s. Matthews (Sligar), "Woolen Trade"; Crockett, *Woolen Industry,* 110.

[4]Alice Gill to Margaret Blythe, 15 January 1882, Jane Gill to Sam and Jessie Blythe, 7 May 1887, WMSHS.

[5]Jane Gill to Sam and Jessie Blythe, 7 May 1887, Caroline Watkins to Mrs. M. D. Scruggs, 9 March 1884, WMSHS.

[6]Rucker, "Watkins Mill in Transition." The historian cannot answer all questions or make neat conclusions from such sources. The exactness of deeds and ledger books does not allow for all the variables of friendly sales within the Watkins family.

[7]"Waltus L. Watkins in acc't with Jno. Watkins & Bros., Sept. 25, 1882," WMSHS; Rucker, "Watkins Mill in Transition."

[8]Neither the structure nor the equipment was listed in any extant documents. It is known that a $25,000 figure was mentioned in negotiations during the summer of 1882, when Fort Worth investors inquired about purchasing the textile firm and moving it to Texas. (After surveying the out-of-date machinery, their interest evaporated.) Rucker, "Watkins Mill in Transition."

[9]Waltus Watkins eulogies by [J. J. Felts?], 3 May 1884, and W. C. Barrett, c. 24 January 1884, Mount Vernon minute book, 1865–87 vol., WMSHS; Waltus Watkins obituaries, Research files, WMSHS; DAR, *Centennial Souvenir,* 117–18; *Portrait and Biographical Record,* 216. The statement in a 1920 history of Clay County that the vacant factory still stood as "a mute witness of the industrial past" included the inference that Waltus Watkins's local successors had not pursued industrialization as intently as he. Woodson, *History of Clay County,* 743–44.

[10]Lyman Wilder memorandum, 31 August 1874, WMSHS; Caldwell, "Watkins Mill", 373-74.

[11]Ogren, *Watkins Mill,* 2–5, 71. Ogren's interpretation of the Watkinses' operations did not take into account the woolen industry in Missouri. Rather, he judged the Watkinses' residentiary midwestern woolen mill against large-scale textile factory operations in the East.

[12]Ogren, *Watkins Mill,* 108–9, 115.

[13]J. Porter to John Watkins, 1 July 1879, WMSHS.

[14]Ogren, *Watkins Mill,* 118–22.

[15]Ogren, *Watkins Mill,* 123–24.

[16]Gross, "Importance of Research," 23–25.

[17]Crockett, *Woolen Industry,* 112.

[18]The advertisements also noted that buyers interested in shorthorn cattle, "pedigrees warranted correct," could call at the mill or correspond through the firm's Lawson address. Advertisements, WMA Collection, WMSHS.

[19]For the 1870 census, Waltus Watkins estimated the value of farm products sold, used, or on hand at $4,000 ($57,958 in 2003 dollars). In 1890, the combined values reported by John, Joe, and Judson Watkins were $7,775 (2003 value $158,552); see Table 1 source notes for 2003 estimates. Population Schedules, Ninth and Tenth U.S. Censuses, 1870, 1880; John H. Watkins notes on Eleventh U.S. Census (1890), WMSHS; Christensen and Kramer, *History of Missouri,* 15, 33, 48.

[20]Rucker, "Watkins Mill"; Dagley, Interview.

[21]Roney, *From Entry Fee,* 23–34.

[22]Crockett, *Woolen Industry,* 126–31; Gross, *Watkins Mill,* 53.

Epilogue

What Eventually Will Become of the Mill?

Interest in the mill as a historic entity emerged almost as soon as it ceased to operate as a business. Through a happy combination of circumstances—primarily its rustic isolation and the owner's inability to sell off its outdated machines—the factory and much of its equipment were still intact when Clay County celebrated its centennial in 1922. The anniversary publication featured two pages on Watkins Mill and observed that "it is the most interesting object from a historical standpoint in Clay County."[1] In a January 1923 interview with the *Kansas City Star,* "Uncle John" Watkins (age 81) stressed that the workplace was exactly as when he closed the factory—not a thread touched on the looms, not a machine dismantled or moved. "Why change it? I like to think of the times gone by. We don't need the revenue the property would bring and I think the mill, in its present condition, is of real historic value."[2] (Such recollections, perpetuating a myth that the factory had closed abruptly and that the machinery was totally intact, show that memory had already begun to overwhelm fact.[3] In truth, as noted earlier, John Watkins had apparently succeeded in selling a fulling mill, and possibly other pieces of equipment as well. In addition, he used the mill building for storage for his farm equipment business. However, there is no question that the mill largely remained as it had been in its final days.) The *Star* article also noted rumors that Henry Ford sought the property "as a historic relic," possibly to be transplanted to Greenfield Village, his emerging museum in Michigan.[4] Left unanswered was the article's summary query: "What eventually will become of the mill?"

It took some years for recognition of the site's historic value to evolve into an active preservation effort. In the meantime, John Watkins's sense that the homestead and factory ought to be preserved was passed to nephew J. Spencer Watkins, who managed the home and grounds for his aging relatives from 1932 until 1945. In July 1934, Spencer orchestrated a large public gathering to honor the eightieth birthday of Aunt Carrie, "a charming picture in her dainty blue and white flowered dress." Though blind and deaf, she personified the "graciousness and hospitality" exhibited by her family for a century. A crowd estimated at two hundred listened to Spencer tell "interesting bits of history concerning the farm and woolen mill to those anxious to add to their knowledge of life" in the county's early days. Among the stories was one shared by older residents "that Frank and Jesse James, who lived nearby, helped themselves to the fancy woolen products of the mill for use as saddle blankets."[5]

Like his uncle, Spencer conducted tours, not only of the home and mill but also of the nearby octagonal school and Baptist church.[6] The Works Progress Administration guide for Missouri, published in 1941, noted that the mill was open daily from nine to five and lay in relative seclusion, accessible by gravel roads and farm lanes.[7] The guide also remarked on an initiative "to interest the state in purchasing the Watkins Farm and maintain it as a museum." The Watkins heirs supported these efforts, which the Clay County Historical Society (CCHS) had begun in 1934, but it would take many more years of work to accomplish the goal.

During World War II, the mill machinery was imperiled. In the first decades of the twentieth century, John Watkins had acted as a dealer for Deering farm machinery. He was also a prominent buyer at many auctions throughout the countryside. Wartime scrap drives found the Watkins grounds heaped with rusting implements, and a reported fifty tons were carted off for the war effort. Ethel Massie Withers, mainstay of the county's history movement, enlisted the aid of then-Senator Harry S. Truman to make certain that machinery inside the mill was not confiscated.[8]

The year 1943 saw a major turning point in the history of the Watkins property as Carrie, the last surviving sibling, went to live with a niece in the far west. Disenchanted with Spencer, fifteen inheritors (who had been waiting since the Great Depression for an upturn in real estate prices) petitioned the Clay County circuit court for partition of the historic tract. In 1944, a buyer for the entire 1,600 acres appeared in the person of Henry Frass Jr. of Texas. It was reported that he had paid $80,000 for the parcel and that he planned "to rehabilitate the property and develop a stock ranch on the acreage and also to perpetuate the historic old Civil War woolen mill" and nearby twenty-room mansion. Several members of the Frass family lived in the Big House over the next thirteen years. During that time, Henry's sister, Emelia Frass Mason, attempted to make the mill and grounds into a showplace for sightseers.[9]

When the 1953 annual meeting of the CCHS convened at the Watkins farm, with an assembly of 250 members and guests, the key agenda item was politics.[10] Though the *Kansas City Times* noted the failure of prior endeavors, the future looked brighter. Local politicians pledged that they would seek state appropriations. Officials on the newly reorganized Missouri State Park Board (MSPB) offered their support. From Ray County came added impetus. Ray Coffman, sometime mayor of Lawson, and Ruth Roney, local historian, worked through the local Rotary Club in support of the state park movement. Roney succinctly asserted the national significance of the mill as "the only one in the United States of its period to be still standing, machinery in place."[11] She feared that the estate would be divided and a worthy preservation opportunity lost forever.

John Watkins Oliver assumed a crucial role in the spring of 1954. The Kansas City attorney, a member of the Watkins family's Ray County branch, was invited to address the CCHS. Local sentiment, more than ever, prayed for state acquisition of the site. The General Assembly for 1953 had doubled membership of the MSPB, placed historic sites under its responsibility, and for the first time appropriated more than a million dollars for parks in the two-year budget.[12] The audience probably anticipated a more optimistic message than he delivered. He was not certain if the local goal ought to be "a large recreational park which would naturally include the historic site" or to remain with "a smaller historical site park."[13] The crux of his remarks was that strong grassroots support would be the key to raising sufficient public funds for the purchase. He concluded, "The work in which you are engaged is of great importance.... A deeper understanding of our own times is made possible only by understanding of our past."

Unfortunately, there were then insufficient state funds to purchase the property, much less finance restoration.[14] The Frass-Mason families determined to put the property up for auction rather than wait further on state officials. In the spring of 1958, a circular from Wayne Cook Associates was directed at farmers and ranchers in the region who wished to acquire property "with fascinating historical buildings and background."[15] In the opinion of the auctioneer, the three parcels of Watkins's property were "highly suitable for farm and ranch, game preserve, recreational center, historic shrine or an investment held for future development." The advertisement noted that "the old woolen mill and contents will be offered separately," and emphasized in three different places that the wool factory "stands just as it was, complete with all its machinery in place." An additional piece of bait for prospective buyers was a "Civil War saddle (said to have been used by Jesse James when his horse was shot out from under him)." Prospective bidders were invited to inspect the property from 21 April until the sale on 6 May.

As a sizeable crowd, estimated at two hundred, assembled at the designated hour for the auction, the destiny of the site hung in the balance. Ranchers, antique hunters, and scrap dealers joined the throng at the front porch of the three-story brick mansion. Among the historically minded was a representative from the Smithsonian Institution, Grace Rogers (Cooper), who was particularly interested in two rare ring-frame ply-twisters.

Two men from Independence, Forest Ingram and Lee Oberholtz, had been unable to make it from their jobs at the Allis-Chalmers farm

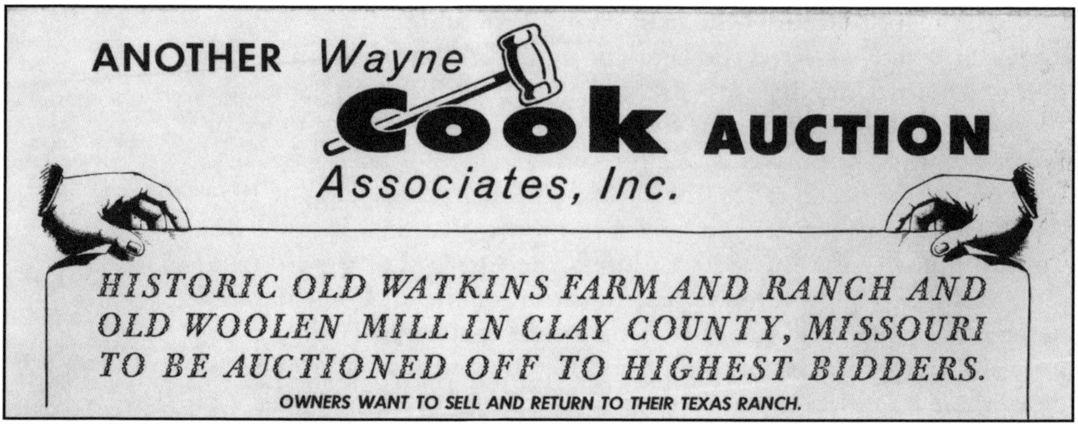

The Watkins Woolen Mill and its entire contents will be sold separate and apart from the rest of the property. It will be sold either in its entirety or piecemeal, furnishing a fine opportunity to dealers, collectors, prop departments for studios, museums, foundations and historical societies to acquire some valuable and interesting items. The purchase of the entire property and mill would furnish someone an excellent home and museum of the early day woolen industry. Foundations, Associations or Machinery Manufacturers are presented herewith an outstanding opportunity to preserve the oldest woolen mill still standing in the U.S.A. with its contents and all machinery. The sale date is 10:00 A.M. (C.S.T.) May 6, 1958, and inspection can be made any time after April 21st. If more information is desired, or questions should arise, contact the auctioneers.

Portions of the circular advertising the auction. DNR, WMSHS.

Cofounders of the WMA, from left: Forest Ingram, Lee Oberholtz, and George Reuland. WMA.

equipment plant in time for the real estate auction. They did, however, find themselves among the thirty-five bidders inside the factory. Together they also represented a colleague, George Reuland, general manager at Allis-Chalmers. In periodic office discussions held since learning of the proposed auction, the three had talked of purchasing pieces of machinery as mementos of bygone days of the American industrial revolution. They had few expectations.[16]

As the mill sale commenced, Ingram and Oberholtz became separated. The impatient auctioneer determined that the contents would be put up for bid as one lot rather than piecemeal. He also, Oberholtz later recalled, seemed partial to any bid that did not originate from a scrap dealer. When the gavel signaled a surprisingly quick end to the bids, Oberholtz was astonished to discover that the successful bidder, at $650, was Ingram.

The auction seemed to signal the end of local efforts at preserving Watkins Mill, as the winning real estate bidder, a Texas rancher, reportedly intended to turn the mill structure into a slaughterhouse or soap factory. The integrity of the site and its machinery, cherished as its distinctive attribute, would surely be destroyed. However, the crisis quickly passed when the Texan—perhaps a "ringer" planted by the auctioneer—failed to muster sufficient down payment. Instead the Watkins property passed to George Stilley, the second highest bidder. A Raytown farmer who had been eyeing the tract since 1945, Stilley quickly put cattle on the grazing land. He assured the Allis-Chalmers associates that the equipment could remain indefinitely in the factory.[17]

May 1958 was a whirlwind. Within two days after the sale, Reuland, Ingram, and Oberholtz decided to form the Watkins Mill

Association (WMA), formalizing the sharing of the costs of their purchase. Stilley visited Reuland at the Allis-Chalmers plant to develop an understanding about the WMA'S tenancy in the mill. On a rotating basis Reuland, Oberholtz, and Ingram would draw on their own personal funds to cover the lease. It was determined that documentary sources found in the Watkins property would be best safeguarded if placed in the newly opened Harry S. Truman Presidential Library in Independence, under the auspices of the Jackson County Historical Society.[18] (These archives were returned to the site in 1987.)

By 10 April 1959, the WMA had incorporated as a not-for-profit organization, with historic preservation goals that included "to procure, preserve, restore, maintain, study and promote the Watkins Mill Estate in Clay County Missouri, and to preserve and make available to the public the historic operations in early industry, community life, church and school as exemplified by the historic old Watkins Mill."[19] While the initial focus would be on stewardship of these sources, a second concern was a strategy to promote tourism and economic development.[20] Reuland was named president, Ingram served as registered agent, and Oberholtz filled the post of secretary. From June 1959 onward, the WMA conducted mill tours at the site, "now making what may be its last stand as a possible historic museum."[21]

George Stilley proved to be a substantial patron of the preservation movement. On 28 June 1960, he sold the mill building and 12.5 acres to the WMA; by the time he died in 1962, he had optioned the other historic structures at the site to the Association. The bulk of his Watkins landholdings were sold to Joe Capps Jr. of Liberty, who was amenable to the goals of the WMA. Capps's father had worked with the CCHS on behalf of the state park effort, and Capps himself volunteered as a tour guide under the WMA.[22]

The efforts of the WMA and the significance of the site drew national attention. Grace Rogers Cooper returned to Watkins Mill in the early summer of 1961. She expressed her professional opinion that the site was "a unique preservation of a 'missing link' in American textile manufacturing.... [A]nyone in America interested in restoration of historical objects would be impressed by the mill." She credited the rural setting as "probably responsible for the unusual state of preservation."[23] The following spring, an illustrated article in the *Museum News* provided a history of the erection of the buildings as well as later efforts at preservation. Author W. Howard Adams concluded, "Watkins Mill offers not only a pleasant and quaint setting for exploring the not-so-distant past, but also it provides an unexplored historical document capable of adding to our understanding of" the Industrial Revolution in nineteenth-century America.[24]

Since the summer of 1960, the WMA had been cultivating support among parks professionals, focusing on a strategy of passing the site from the WMA to the county and eventually to the state. Fearing the encroachment of real estate developers on the Watkins property, WMA supporters argued that the region would be better served by keeping the tract "intact as a state or national park shrine."[25] As 1962 commenced, the WMA more than ever aimed at state park status. A special meeting in February sought alliances between the WMA and community leaders in Clay County. Three elements favorable to the association's goals now coalesced. First, better highways (beginning with the improvement of U.S. 69 and then construction of Interstate 35) would enhance public travel to the site and augment demands for more recreational space. Second, the Kansas City newspapers, in editorials, feature stories, and reports, gave increased support to state acquisition. The *Star* set forth the political agenda: "While the state playgrounds have been expanding ambitiously in other sections of Missouri, the Kansas City metropolitan area has been all but bypassed.... [I]mportant as the mill could become as a shrine, it would in no way replace the need for a major state park in this part of Missouri." There was a need for "a total recreation pattern for Western Missouri" to satisfy the increasing leisure-time demands of metropolitan Kansas City.[26]

Third, more astute tactics came to the fore. In 1961, George Reuland had commenced parleys between the WMA and the MSPB. He knew that the envisioned park would have to encompass sufficient acreage and that the land would have to be donated, not sold, to the MSPB. William Dye, the managing editor of the *Lexington Advertiser News,* spearhead of the acquisition movement, and a member of the MSPB, wrote to Reuland, "It is most important that we strive to get more property to give the state. The board's decision probably will hinge on this."[27] In the spring of 1962, the WMA orchestrated a more enthusiastic and broader-based campaign than any prior attempt. With a site hearing scheduled for 4–5 May, endorsements from experts and swelling civic enthusiasm were directed toward the MSPB.[28] To aid the MSPB in its deliberations, Robert L. Elgin of Recreation Associates was hired to visit the site on 28–30 May, hosted by Forest Ingram. In his report to Director of Parks Joseph Jaeger Jr., Elgin made three key points:

- The mill was "on a par with the best of these Eastern plants [New England's historic mills] when judged from the viewpoint of being historically important and mechanically complete."
- Itemized estimates of repairs needed for the mill, house, several outbuildings, church, and school totaled $156,369.
- The minimum goal for acreage was placed at 460, with recommendations that tracts to the south and east of the mill site be acquired.[29]

Elgin's acreage recommendation became the first of the conditions placed by the MSPB on an acceptance of the Watkins Mill site. The *Kansas City Star* suggested in an editorial that the necessary local funds—a "modest" sum of $115,000—could be gained either by a county bond issue or by a private fund campaign.[30]

From the WMA's point of view, there was no alternative but a bond issue. They worked with county officials to develop a fact sheet outlining the $232,000 request for an estimated 900 acres.[31] Although they also circulated a flyer emphasizing the historical significance of the site,[32] the primary objective had become establishment of a substantial park for western Missouri. The supporters' pitch was that Watkins Mill State Park would offer something for everyone: rest and recreation as well as a glimpse of history. Furthermore, Clay Countians' state tax dollars would return home, and tourists would add to the local economy, at an estimated cost of $5 each for the "vast majority" of county property owners. The *Kansas City Times* ran not only a supportive editorial but also an accompanying feature story on the WMA. The paper touted the property as "West Missouri's first major state park" and pointed out that "Clay County would receive a major playground and historic shrine." The story concluded that in the November general election, "Watkins mill is facing its greatest test for survival in its 101-year history."[33]

The bond issue did receive a majority (52.5 percent) of favorable votes but fell short of the 66.7 percent required for passage. The day after the election, Al Bohling from the Kansas City newspapers encouraged the WMA to quickly mount another drive that could separate the issue from other questions on the autumn ballot.[34] Representatives of the association met with the Clay County Court, and a new campaign committee was formed, with a revised goal of $184,000 for 780 acres.[35] Another element of urgency was added: "Clay County is now competing with Jackson County for a State Park.... In simple terms if Clay County doesn't want a State Park, Jackson County will get it."[36]

In the second campaign, the preservation of Watkins Mill was promoted as contributing to the quality of life of citizens in western Missouri rather than as an end in itself. Later that summer, a national parks historian summarized the fundamental strategy: "It is primarily the mill which provides the area with a uniqueness.... [Although it] is the mill that justified the creation of the park...most visitors will be interested primarily in picnicking in the recreational aspects of the area; to these, the visit to the mill and the historical section of the area, will be only incidental."[37] A week before the election, the *Star* ran an editorial called "The Last Chance to Save Watkins Mill." The writer noted that "prospects of approval seemed brighter" and opined that with passage of the bond issue, "Clay County will receive something of a bonanza." (A midwestern "miniature Williamsburg" was one of the projected scenarios.) The article's summary sounded a somewhat revised refrain from the past four decades: "The loss of this major park and shrine would be a recreational tragedy for the region and the state."[38]

When the ballots were counted on March 19, nearly 71 percent of the votes countywide were "Yes." Gallatin, the most populated and urbanized but also most distant township, had voted down the first bond issue; this time 76.8 percent of its votes were affirmative (perhaps indicating the effectiveness of the WMA's regional appeals).[39] John Watkins Oliver's 1954 advice—"Strong citizen support is the key.... Intelligent planning and hard work are the only methods that I know about that will place that key in your hands"—had proved on target.[40] The *Star*'s editors observed, "The people who worked so long and so hard to save Watkins Mill deserve the gratitude of the entire state."[41] An estimated four hundred bond-drive workers picnicked together at the site on 23 June to celebrate their victory.

Two bond drives in five months did not interrupt the major curatorial mission of the WMA.[42] As 1963 wore on, Jack Simpson and associates in Cameron mustered funds to publish *Smokey Row,* a highly fictional piece by Bertha Bell Entrikin; Simpson even proposed that Walt Disney be enlisted to make a film at the site.[43] There was impatience concerning lagging land surveys and warranty deeds. As late as the last day of the year, transactions were still being made by the WMA.[44] The state took possession of the site on 1 January 1964. Restoration work began, with the WMA accepting responsibility for cleaning and repairing the mill machinery. By 1 November 1965, the mill was ready to reopen, with the WMA under contract to operate the site properties.

National recognition quickly followed. In late fall of 1966, word came that the mill, house, and outbuildings had been named a Registered National Historic Landmark. At the dedication on 23 April 1967, a commemorative plaque was placed on a large rock that George Stilley had moved to the location before his death, and featured speaker Edward Hummel emphasized the uniqueness of the landmark: "There is only one mid-nineteenth century textile mill with all its original machinery and it is Watkins mill." A 1969 article in the journal of the Society for the History of Technology claimed that "Watkins Mill is the most important industrial relic for the study of textile history in this country, and probably the world.... Watkins Mill as a whole is such a straightforward presentation, with no effort to seem more antique than it is.... Only the noise and the smell are missing." Then on 15 April 1980, the American Society of Mechanical Engineers designated the Watkins Woolen Mill as a National Historic Mechanical Engineering Landmark.[45]

Transition in ownership of Watkins Mill from the WMA to the State of Missouri meant that professionals in the fields of historic preservation, restoration, and interpretation could augment the prior efforts of the Watkins family and the WMA, with state funds available for at least some of the work. A series of archaeological digs commenced immediately under the supervision of Booker Rucker.[46] In

1978, scholars Laurence Gross and Paul Ogren spent the summer in Missouri working for the Historic American Engineering Record. While Gross focused his attention on the buildings and the machinery of the woolen industry, Ogren investigated correspondence, business records, and other documentary remains of the Watkinses' endeavors. Their findings were complemented by the work of noted photographer Jet Lowe. Because of their joint efforts, the Library of Congress gained an invaluable photograph collection of the Watkins Mill machinery as well as interpretations of artifacts and business practices.[47]

As the State of Missouri has sought to preserve, restore, and interpret Watkins Mill, the focus has been on re-creating the household and agricultural practices of the 1870s. Not only was that decade the peak of the mill and farm productions, it was also captured in the artistic renderings in the 1877 *Illustrated Historical Atlas of Clay County.* Today's living history programs demonstrate such household processes as canning and preserving fruits, making soap, and quilting. Heirloom gardening projects use techniques of the period to raise vintage plants. Similar efforts are made in animal husbandry, focusing on the sheep herd and poultry.[48]

The industrial components at the site—the mill building, the power source and train, and the textile machinery—are not included in demonstrations. Even if a modern power source were provided and all the machinery restored to running order, the costs—both financial and historical—would be prohibitive. Site administrator Ann Matthews (Sligar), now retired, has summarized the issues: "There is no way [Watkins Mill] using ancient equipment, hard to find supplies, and labor intensive processes could make it.... All of these things, training, realism, economics and safety are very real considerations. However, our real reason for not operating the mill is the value of the machinery.... At one time there were 881 woolen mills of this type in the Midwest region. This is the only one where such evidence still exists." In other words, while demonstrations with actual equipment would be instructive and entertaining, the inevitable result—deterioration and loss of irreplaceable artifacts—could not be justified.[49]

The state decided to interpret the factory operations by alternative means. A $1.2 million Visitor Center and Museum opened in May 1992; the first complete set of exhibits and photographs was unveiled in December 1996. Through audiovisual presentations and guided walking tours, visitors are exposed to the world of the Watkins operatives and the effects of industrialism on Missourians a century ago. In recognition of the quality of interpretation, the Society for the History of Technology in 1999 named the site co-recipient of its annual Dibner Award for Excellence.[50]

The contrasts between past and present are clearly evident to Watkins Mill visitors. Those coming from metropolitan Kansas City in their automobiles face a deceleration in speed limits: 70 mph on the interstate highway; 55 mph on State Route 92, 20 mph along the park's paved road, and ultimately a walk on the Lane from the Visitor Center to the site's buildings. Nearby residential development lures city dwellers to middle-class housing, large yards, and small-town ambience. The modern equivalent to Bethany is a 5-acre farm with three horses for recreational riding, while the modern successor to Waltus Watkins's commercial enterprises is a pasta-making firm in Excelsior Springs, one of the largest in the nation.[51]

The Watkins Mill Association continues to supply funds for the site's special projects. The WMA membership rolls have been depleted by deaths or retirements, yet enthusiasm still drives the local stalwarts and supporters such as David Ross and Hal Sandy. A mid-1980s accolade to the men and women who had worked so hard on behalf of the Watkins legacy is still pertinent today: "There is much work to be done still, but the Watkins Mill Historic Site is a marvelous preservation of factory and farm life in the 19th century.... It's the kind of realism no amount of money can buy. The state and the country owe a tremendous debt for this historic site's glimpses at an era of this country's past glory."[52]

NOTES TO EPILOGUE

[1]DAR, *Centennial Souvenir,* 117–18. To place the local preservation movement within the national context, see Kammen, *Mystic Chords.* His interpretative emphasis is that "in the United States, more often than not, memory has served as a bulwark for social and political stability—a means to valorizing resistance to change" (p. 59). He notes "the emergence of an American aesthetic, centering on simplicity, and a passion for Americana that swept the nation during the 1920s; the partisanship promoted by regionalism" (p. 307).

[2]"The Mill That Waits for the Miller," *Kansas City Star,* 28 January 1923; Matthews (Sligar), Interview by Needham.

[3]This trend was reinforced by the apparent failure of those second generation Watkinses who spent the large part of their lives at the location—John, Joe, Judson and Carrie—to commit their recollections about the wool factory to the historical record. They left behind only a few family photographs and a bit of oral tradition. The third generation, more distant from the mill both spatially and emotionally, came to rely on myth and misinformation. It has been the fourth and subsequent generations, such as Chad Means of Chicago and Byran and Betty Frass McDaniel of suburban Kansas City, who have curiosity about Watkins ancestors, hold family reunions, and proffer artifacts, manuscripts, and oral traditions to the site's collections.

[4]There is no documentary proof now available at Watkins Mill or Greenfield Village of Henry Ford's reputed interest in the mill. The claim is only one of several questionable points made in the article. Matthews (Sligar), Interview by Needham; Oberholtz, Interview by Needham.

[5]*Liberty Chronicle,* 2–5 July 1934.

[6]Watkins, *Lucky Montana Cowpoke,* 14; "Notes and Comments," *Missouri Historical Review* 33 (1938–39): 422–23; Ethel Massie Withers, "The Old Watkins Mill May Become a Museum in a State Park," *Kansas City Star,* 28 July 1940.

[7]"A Rich Harvest of Walnut Comes from Historic Farm," *Kansas City Star,* 13 November 1937; "Why Not Those Who Established Industry," *Liberty Advance,* 13 February 1939; "The Old Watkins Mill May Become a Museum in A State Park," *Kansas City Star,* 28 July 1940; *Missouri: A Guide,* 513. The efforts of the CCHS on behalf of the Watkins properties

are chronicled in "Notes and Comments," *Missouri Historical Review* 33 (1938-39): 422-23; 34 (1939-40): 250-51, 542; 35 (1940-41): 114-15, 133-34; and 38 (1943-44): 480.

[8]"Scrap Iron Goes to War at Watkins Farm," *Kansas City Times,* 16 September 1942; "Clay County Would Make a State Park," *Kansas City Times,* 3 October 1953; Ethel Withers to John Watkins Oliver, 3 March 1954, WMA Collection, WMSHS. The legend of Harry Truman's involvement is recounted in "Long Fight to Save Historic Mill Near Climax," *Kansas City Times,* 3 November 1962. A survey of holdings at the Harry S. Truman Presidential Library in Independence yielded no pertinent documentation of this episode. Mrs. Withers's son has no clear recollection of the incident (but does commend his mother's veracity). Conn Withers to Louis Potts, 18 March 1988, Western Historical Manuscripts Collection, University of Missouri-Kansas City.

[9]"Pre-Civil War Setting for Historical Society Dinner," *Excelsior Springs Daily Standard,* 7 March 1945; "Historical Group Barbeque at Old Watkins Mill Farm," *Liberty Tribune,* 21 September 1953. Inaccuracies in this article included the assertion, possibly traced to Ruth Roney, that the old portrait of Waltus Watkins in the mansion was the work of George Caleb Bingham.

[10]"Clay County Would Make a State Park," *Kansas City Times,* 3 October 1953; enclosures in Ray C. Coffman to John Watkins Oliver, 8 March 1954, Ray Coffman Papers, WMSHS.

[11]Roney, "Data."

[12]Rucker, Interview by Potts; C. W. Boutin to John W. Oliver, 1 March 1954, WMA Collection, WMSHS.

[13]John Watkins Oliver, speech, 15 March 1954, WMA Collection, WMSHS.

[14]Flader, *Exploring Missouri's Legacy,* 1-23. In retrospect, the years 1955 through 1970 would be seen as a high point of the state's response to a widespread recreational boom. Joseph Jaeger, a conservation forester, served in the period as an astute and dynamic steward of the Parks Department. He would add twenty-six parks to the state system; nearly all would be donated. "Like the Watkins Mill—Members State Park Board Tour the Grounds," *Kansas City Star,* 12 December 1954; Mason scrapbook, WMSHS.

[15]Ray Coffman Papers, WMSHS. The advertisement's text relied heavily upon Roney, "Data."

[16]Oberholtz, Interviews by Potts, 23 June 1987, 29 February 1988, and 15 April 1988; Oberholtz to George Reuland, 9 May 1958, WMA Collection, WMSHS.

[17]Stilley paid approximately $65 per acre for the 1,600 acres and buildings. His bid was $168,000. Oberholtz, Interview by Potts, 23 June 1987.

[18]From the initial auction, Ingram claimed rights to a surveyor's compass and tripod, which he returned to the site later. Oberholtz, Interviews by Potts, 29 February 1988 and 15 April 1988; L. E. Oberholtz to George Reuland, 9 May 1958, WMA Collection, WMSHS; Russell, "Watkins Mill Association."

[19]Oberholtz, Interviews by Potts, 23 June 1987 and 15 April 1988; "Independence Men Stop Loss of Historical Treasure," *Kansas City Star,* 14 July 1982; L. E. Oberholtz to George Reuland, 9 May 1958, WMA Collection, WMSHS; WMA bylaws, 1959, Ray Coffman Papers, WMSHS.

[20]Oberholtz, Interviews by Potts, 23 June 1987 and 15 April 1988.

[21]"Visiting Clay County Woolen Mill, Intact After 100 Years," *Kansas City Star,* 7 June 1959.

[22]L. E. Oberholtz to WMA Trustees, 23 and 30 December 1962, WMA Collection, WMSHS; "Mile Posts Along the Way," c. 1967, WMA Collection, WMSHS; Oberholtz, Interviews by Potts, 23 June 1987 and 29 February 1988; "Historic Watkins Mill to Continue as a Museum," *Kansas City Star,* 17 December 1969. The Association in the 1959-61 period got curators from the Nelson-Atkins Museum in Kansas City to do an appraisal of the mill's artifacts. The estimate was $27,910. Russell, "Watkins Mill Association," 5.

[23]"Smithsonian Eyes Historical Mill with Envy," *Kansas City Star,* 2 June 1961.

[24]Adams, "Story of Watkins Mill," 20-22.

[25]Rucker, Interview by Potts; "Watkins Mill Possibilities," *Excelsior Springs Shopper,* 2 June 1960; L. E. Oberholtz to Association Members, 26 September 1960, WMA Collection, WMSHS.

[26]"The Watkins Mill Merits Preservation," *Kansas City Star,* 22 February 1962. The movement in Clay County was part of a phenomenon nationwide, as "conservationists, environmentalists and historic preservationists fashioned an emerging alliance under the banner of heritage preservation." In Clay County, preservationists allied with state recreation advocates. Wallace, "Reflections," 185.

[27]William Dye to George Reuland, 7 March 1962, Reuland to Dye, 9 April 1962, WMA Collection, WMSHS.

[28]William J. Murtagh to Joseph Jaeger Jr., 7 June 1962, WMA Collection, WMSHS; "City, C. Of C., Back Mill Park Project," *Kansas City Times,* 5 May 1962.

[29]R. L. Elgin to Joseph Jaeger Jr., 7 June 1962, WMA Collection, WMSHS; Rucker, Interview by Potts. The MSPB "raised the ante twice more for some 700 acres" before the election. In addition, local conferences had evidently determined that the bigger the recreational area, the more attractive the site would be to the local voters and the state. "Mile Posts" [1988 addendum], WMSHS; Oberholtz, Interview by Potts, 23 June 1987.

[30]Joseph Jaeger Jr. to George Reuland, 6 June 1962, WMA Collection, WMSHS; "Decisions Within Two Weeks On Course for Watkins Mill," *North Kansas City Dispatch,* 10 July 1962; "Fine Area Park around Watkins Mill," *Kansas City Star,* 5 July 1962.

[31]"Our Opportunity: A State Park for Clay County," broadside, [16 September 1962?], WMA Collection, WMSHS; "Fact Sheet on Proposed Watkins Mill State Park Bond Issue," [14 December 1962], WMA Collection, WMSHS.

[32]"Our Opportunity," WMSHS. This poster, circulated county-wide, proclaimed, "REWARDS: RECREATION, HISTORY, REVENUE."

[33]"Mile Posts," WMSHS; *Town and Country Leader,* 26 October 1962; "The Watkins Mill Should Be Saved" and "Long Fight To Save Historic Mill Near Climax," *Kansas City Times,* 3 November 1962.

[34]Watkins Mill Committee for State Park Bonds, 16 January 1963, WMA Collection, WMSHS.

[35]Oberholtz, Interview by Potts, 23 June 1987.

[36]"Fact Sheet," WMSHS; Oberholtz, Interview by Potts, 15 April 1988.

[37]Roy Mattson to Joseph Jaeger, 9–10 July 1963, WMA Collection, WMSHS.

[38]Oberholtz, Interviews by Potts, 23 and 25 June 1987; "The Last Chance to Save Watkins Mill," *Kansas City Star,* 12 March 1963.

[39]At 7,488 voters, the spring turnout was only 41 percent of the previous election. Comparison of 1962 and 1963 returns can be gained from "Committee Comments," [16 January 1963], WMA Collection,WMSHS; and "Watkins Mill Bonds Passed," *Kansas City Times,* 20 March 1963.

[40]John Watkins Oliver speech, 15 March 1954, WMA Collection, WMSHS.

[41]"From An Old Mill, A Fine Park Will Grow," *Kansas City Star,* 20 March 1963. See Joseph Jaeger to Governor John M. Dalton, 13 April 1963 (Dalton Papers), for emphasis on need to provide "a state park near the greater Kansas City area."

[42]Oberholtz, Interviews by Potts, 23 and 24 June 1987; "Watkins Mill News," 19 September 1963, WMA Collection, WMSHS.

[43]Entrikin, *Smokey Row;* Jack V. Simpson to Ray Coffman, 8 June 1963, WMSHS.

[44]Oberholtz, Interview by Potts, 25 June 1987; Oberholtz to Reuland, 23 and 30 December 1963, WMA Collection, WMSHS; Russell, "Watkins Mill Association," 6.

[45]Hummel also made the point that the local and statewide preservation movements mirrored cresting interests nationwide. Oberholtz, Interview by Potts, 25 June 1987; "Dedicate Mill as Historic Site," *Kansas City Star,* 24 April 1967; Hippen, "Industrial Textile Machinery."

[46]Rucker, "Archaeological Investigations"; Bray, *Archaeological Investigations.* Rucker went on the become supervisor of all Missouri state historic sites; he retired in 2002.

[47]Gross, *Watkins Mill;* Ogren, *Watkins Mill;* Gross, "Importance of Research"; "Watkins Mill," HABS/HAER.

[48]*Illustrated Historical Atlas.* See also Sligar, *Waltus Watkins,* 102–15.

[49]Matthews (Sligar), "Why We Decided."

[50]Flader, *Exploring Missouri's Legacy,* 328–33; *Missouri Resources* 14 (Spring 1997), 18.

[51]"Area Pasta-Maker Agrees to Supply Mueller's Brand," *Kansas City Star,* 15 April 1997; "Pasta Firm Buyout Halted," *Kansas City Star,* 28 March 1991.

[52]"Independence Men Stop Loss of Historical Treasures" and "Window on the Past," *Kansas City Star,* 14 July 1982.

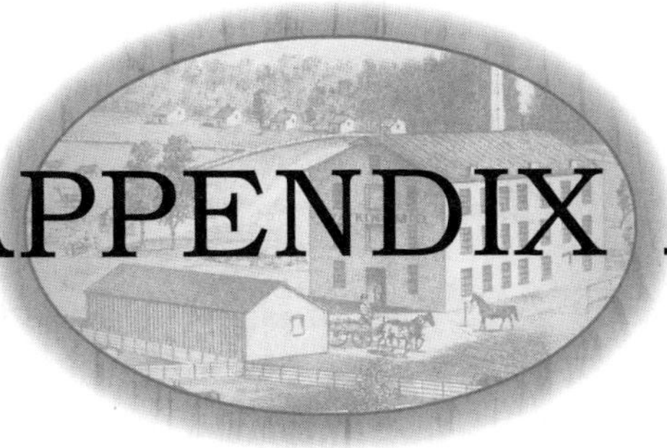

APPENDIX A

WATKINS MILL CHRONOLOGY

1806	Birth of Walthall "Waltus" Lockett Watkins in Kentucky
1817	Birth of Mary Ann Holloway in Kentucky
1822	Clay County, Missouri, established
1824	Start of Waltus's apprenticeships with his uncles in Kentucky
1825	Waltus trains as a mechanician (machinist)
1826	Waltus becomes a master of textile manufacturing
1828	Waltus employed in Dudley's spinning mill, Frankfort, Kentucky
1830	Death of Waltus's father, Benjamin Watkins Waltus visits Clay County, Missouri
1831–39	Waltus establishes enterprises in Liberty, Missouri: cotton spinning, wool carding, tannery, and barkmill
1834	Marriage of Waltus and Mary Ann
1835	Birth of John Holloway Watkins
[1836]	Franklin subscription school (later Franklin Academy) organized
1836–c. 1850	Watkins kin move to western Missouri from Kentucky
1837–44	Economic depression
1837	Birth of George Spencer Watkins

1838	Waltus appointed Justice of the Peace for Liberty Township Waltus makes initial land purchases in Washington Township
1839	Waltus's mill in Liberty burns; property sold Watkinses move to Washington Township Birth of Alfred Watkins First Watkins cabin erected Death of John Holloway Watkins
1840	Death of Alfred Watkins
1841	Birth of John Handy Watkins
1842	Franklin Debate Society established
1843	Franklin Library Association established
1844	Waltus appointed to school board Birth of Catherine Jane (Kate) Watkins
1844–48	Waltus serves as postmaster at Bethany
1845	First brick kiln erected on property
1847	Birth of Martha Ann (Mattie) Watkins
1848–49, 1853	Waltus acquires lands in Ray County
1849	Birth of Mary Elizabeth (Lizzie) Watkins Waltus experiences religious conversion; joins Baptist Association at New Hope Church
1840s–50s	Commercial farming and stock raising at Bethany; enterprises include flour and gristmill (c. 1848), sawmill (1850), and icehouse (1856)
1850–c. 1870	Waltus serves as trustee of William Jewell College
1850–54	Brick house and outbuildings built at Bethany
1852	Birth of Waltus Jewell Watkins
1854	Birth of Caroline Emma (Carrie) Watkins

1856	Octagonal brick structure built for Franklin Academy Birth of Adoniram Judson (A. J.) Watkins
1857	Economic depression Mount Vernon Missionary Baptist Church established
1859	Birth of Joe Berry Watkins
1861	Woolen mill in full production (built c. 1859; carding and fulling services possibly offered 1859–60)
1862	Federal troops ransack Bethany Big House Death of Jane Minter Watkins
1863	Carding and spinning expanded; broadlooms acquired
1864	Bushwhackers raid mill
1868	Mill modifications and improvements W. L. Watkins and Son established
1870	Death of Catherine Jane (Watkins) Atchison
1871	Brick Mount Vernon church built
1873–79	Economic depression
1875	Grasshopper plague
1876	Mill modifications and improvements
1882	Waltus ill; deeds mill and lands to John, Joe, and Judson John Watkins and Brothers established
1884	Death of Waltus L. Watkins
1886	Fabric production ceases
1888	John buys brothers' shares in factory
1892	Yarn production ceases
1893–97	Economic depression

1896	Death of Mary Ann Watkins
1898	John closes woolen factory Death of Mary Elizabeth (Watkins) Atchison
1904	Mill's steam engine (used for gristmill) stops working; factory closes for good
1909–30	Annual meetings of Missouri Fox Hunters' Association on site
1925	Death of Waltus Jewell Watkins
1929	Death of Martha Ann (Watkins) Scruggs
1931	Deaths of George Spencer Watkins and John Handy Watkins
1933	Death of Joe Berry Watkins
1934	Death of Adoniram Judson Watkins
1943	Carrie Watkins moves from Bethany
1944	Henry Frass Jr. purchases mill property; moves into Big House with sister Emelia Mason
1946	Death of Caroline Emma Watkins
1958	Property auction; George Stilley gains title Forest Ingram, Lee Oberholtz and George Reuland buy mill machinery
1959	Watkins Mill Association incorporated Mill open for tours
1960	Stilley sells mill building and acreage to Watkins Mill Association
1962	Unsuccessful bond issue for Clay County to purchase 900-acre site
1963	Successful bond issue for purchase of 783-acre site by Clay County
1964	Site donated to State of Missouri; restorations and reconstructions begin

1966	Site designated a National Historic Landmark
1967	Exterior of factory restored
1980	Site designated a National Historic Mechanical Engineering Landmark
1981–84	Major restorations: house, school, church, gristmill, outbuildings
1992	Visitor Center and Museum dedicated

Appendix B

Abbreviated Family Tree

Benjamin Watkins - - - - - - married Jane Minter, April 21, 1801
b. October 1, 1775 — *Chesterfield County, Virginia*
d. June 13, 1830 — *Woodford County, Kentucky*

Jane Minter Watkins
b. March 1, 1781 — *Pittsylvania County, Virginia*
d. August 20, 1862 — *Clay County, Missouri*

Children of Benjamin and Jane:

1. Mary (Polley) - - - - - - married John G. Handy
   b. October 30, 1802 — *Woodford County, Kentucky*
   d. August 6, 1877 — *Mercer County, Kentucky*

2. Martha L. - - - - - - married Abraham B. Luke
   b. September 24, 1804 — *Woodford County, Kentucky*
   d. March 19, 1838 — *Woodford County, Kentucky*

3. Walthall Lockett (Waltus or Walter) - - - - - - married Mary Ann Holloway, March 4, 1834
   b. October 30, 1806 — *Woodford County, Kentucky*
   d. January 24, 1884 — *Clay County, Missouri*

   Mary Ann Holloway
   b. January 5, 1817 — *Jessamine County, Kentucky*
   d. September 24, 1896 — *Clay County, Missouri*

   Children of Waltus and Mary Ann:

   1. John Holloway
      b. January 3, 1835 — *d. November 13, 1839*
   2. George Spencer
      b. February 14, 1837 — *d. February 24, 1931*
   3. Alfred
      b. September 6, 1839 — *d. March 6, 1840*
   4. John Handy
      b. September 16, 1841 — *d. February 9, 1931*
   5. Catherine Jane (Kate) - - - - - - married Dr. John (Hugh) Atchison
      b. August 12, 1844 — *d. October 8, 1870*

6. Martha Ann (Mattie) - married Milus Dickey Scruggs
b. March 7, 1847 *d. May 30, 1929*
7. Mary Elizabeth (Lizzie)- -married William Atchison Jr.
b. September 29, 1849 *d. October 24, 1898*
8. Waltus Jewell - married Minnie Brown
b. March 5, 1852 *d. July 17, 1925*
9. Caroline Emma (Carrie)
b. July 1, 1854 *d. July 6, 1946*
10. Adoniram Judson (A.J.) - married Jane Smith
b. December 24, 1856 *d. July 20, 1934*
11. Joe Berry- married Ina Gooddard
b. April 21, 1859 *d. May 27, 1933*

4. Benjamin Franklin -married Elvira H. Atkins; Anna Jerusha MacArthur
b. May 22, 1808 *Woodford County, Kentucky*
d. September 12, 1871 *St. Louis County, Missouri*

5. Jane M. -married Rev. Philip E. Gill
b. November 30, 1809 *Woodford County, Kentucky*
d. April 3, 1895 *Liberty, Clay County, Missouri*

6. James Minter - married Martha Scearce
b. April 16, 1811 *Woodford County, Kentucky*
d. April 23, 1876 *Clay County, Missouri*

7. Caroline W. - married Sen. Parker Calhoun Hardin
b. May 7, 1813 *Woodford County, Kentucky*
d. July 13, 1847 *Columbia, Adair County, Kentucky*

8. Elizabeth Ann - married Isaac G. Carter
b. June 8, 1814 *Woodford County, Kentucky*
d. October 13, 1858 *Clay County, Missouri*

9. Tabetha P. (Tabitha) - married John W. Gill
b. December 6, 1815 *Woodford County, Kentucky*
d. March 1, 1882 *Clay County, Missouri*

10. Ann Maria
b. May 21, 1817 *Woodford County, Kentucky*
d. July 29, 1838 *Woodford County, Kentucky*

11. Margaret Tolar- married Milton J. Byram; Arnold Duncan Blythe
b. April 10, 1819 *Woodford County, Kentucky*
d. April 2, 1903 *Platte County, Missouri*

12. Rebekah P. (Rebecca) -married William L. Yates
b. May 10, 1821 *Woodford County, Kentucky*
d. November 25, 1885 *Clay County, Missouri*

13. Susan M. -married John R. Hardin
b. August 12, 1823 *Woodford County, Kentucky*
d. October 25, 1850 *Weston, Platte County, Missouri*

APPENDIX C

HIERARCHY OF WATKINS MILL EMPLOYEES, 1870

Although no structural outline of the hierarchy in Watkins Mill exists, one can be reconstituted using 1870 census data, the most fully documented year for Watkins and his employees.[1]

OCCUPATION	PERSONNEL	AGE	SELF VALUE ($)
Farmer and manufacturer	Waltus L. Watkins	63	142,000
	John H. Watkins	28	
Superintendent	James M. Griffin	43	2,000
	John S. Quaeland	30	500
Engineer and machinist	Jesse Walker	54	1,000
Fireman	Elias Goodwin	26	500
Dyer and bookkeeper	William Waers	47	2,000
Dyer	Edward Townsal	40	
Store clerk	Jonathan Cook	67	2,000
Salesman	William Taylor		
Carder	Thomas McGerty	28	800
	John Nailer	25	
Spinner	Michael Murphy	21	
	John McCullough	21	
Wool picker	Doctor F. Marley	36	700
Fuller	Gil Jackson	40	2,200
Miller	William Reed	57	
Weaver	Isabelle Griffin	22	
	Bettie Hulls		
	Arthur Taylor		
	Frances Vermillion		

OCCUPATION	PERSONNEL	AGE	SELF VALUE ($)
Mill worker	Mary Smith		
	Hanna Donohoe	55	300
	Eli Burgess		
Mill hand	James Gromack		
	T. J. Cavender		
	Elisha Cravens		
	Lewis Crowley		
	Peter Cold[2]		
	James Dalzell		
	John Donohoe		
	Dave Estes	40	600
	Rebeun Fessler		
	Jesse Fore	54	1,000
	Jeff Griffin		
	Gideon Harwick		
	Avery Harwick		
	John James		
	Samuel McSpadden		
	Henry Marley		
	John H. Marley	13	
	Joseph S. Marley		
	William A. Marley	16	
	John Nokes		
	Henry O'Dell		
	John Night		
	Charles Peters	28	200
	Adolph Sueflenberg	21	700
	Samuel Walters	17	1,500
	John Waers	13	
	George Washty		
	Sam Waters		

[1]Ninth Census of the United States, 1870, population schedule, Washington Township, Clay County, MO, WMSHS.

[2]Possibly the same as Charles Peters.

Bibliography

Adams, W. Howard. "The Story of Watkins Mill." *Museum News* 40 (May 1962): 20–27.

Archaeological Investigations, Watkins Mill State Historic Site. Vol. 2, *Waers Cabin.* Draft report prepared by Environmental Research Center of Missouri, Inc., May 1990. WMSHS.

Archer, Samuel. "Wool Growing in Missouri." *Bulletin of the National Association of Wool Manufacturers* 10 (1880): 190–95.

Bagg, Sharon. "Kate of Watkins Mill: Letters from a Missouri Plantation in the 1860s." Unpublished paper, University of Missouri–Kansas City, 1991. WMSHS.

Baird, Charles W. *History of the Huguenot Emigration to America.* 2 vols. 1885. Reprint, Baltimore, MD: Regional Publishing, 1966.

Barney, William L. *The Passage of the Republic: An Interdisciplinary History of Nineteenth-Century America.* Lexington, MA: Heath, 1987.

Bender, Thomas. *Community and Social Change in America.* New Brunswick, NJ: Rutgers University Press, 1978.

——. *Toward an Urban Vision: Ideas and Institutions in Nineteenth-Century America.* Lexington: University Press of Kentucky, 1975.

Benson, Susan Porter, Steven Brier, and Roy Rosenzweig, eds. *Presenting the Past: Essays on History and the Public.* Philadelphia: Temple University Press, 1986.

Bloch, E. Maurice. *The Paintings of George Caleb Bingham: A catalogue raisonné.* Columbia: University of Missouri Press, 1986.

Boorstin, Daniel J. *The Republic of Technology.* New York: Harper & Row, 1978.

Bowen, Don R. "Guerrilla War in Western Missouri 1862–1865: Historical Extensions of the Relative Deprivation Hypothesis." *Comparative Studies in Society and History* 19 (1977): 30–51.

Bray, Robert T. *Archaeological Investigations (1981) of First Cabin at Watkins Mill State Historic Site.* Unpublished report, August 1982. WMSHS.

Bray, Robert T. and Jeffrey K. Yelton. *Archaeological Investigations at Watkins Woolen Mill State Historic Site.* Unpublished report, 1990. WMSHS.

Brown, Richard D. *Modernization: The Transformation of American Life 1600–1865.* New York: Hill and Wang, 1976.

Bruchey, Stuart. *Enterprise: The Dynamic Economy of a Free People.* Cambridge, MA: Harvard University Press, 1990.

Bushman, Katherine Gentry, comp. *Index of the First Platbook of Clay County, Missouri: 1819–1875.* N.p.: 1967. Clay County Archives, Liberty, MO.

Bushman, Richard L. *The Refinement of America: Persons, Houses, Cities.* New York: Knopf, 1992.

Caldwell, Dorothy J. "Missouri's National Historic Landmarks. Part 3: Watkins Mill." *Missouri Historical Review* 63 (April 1969): 364–77.

Christensen, Lawrence O., and Gary R. Kremer. *History of Missouri, 1875 to 1919.* Vol. 4 of *History of Missouri,* edited by William E. Parrish. Columbia: University of Missouri Press, 1997.

Clark, Christoper, and Nancy A. Hewitt. *Who Built America: Working People and the Nation's Economy, Politics, Culture, and Society.* Vol. 1. *From Conquest and Colonization through Reconstruction and the Great Uprising of 1877,* 2nd ed. New York: Worth, 2000.

Clark, Patrick. "Breeding Practices of Jack Stock and Use of Mules on Bethany Plantation." Unpublished paper, University of Missouri–Kansas City, 1992. WMSHS.

Clay County Archives, Liberty, MO. Includes Circuit Court Records, Clay County Court Records, Clay County Commission Records.

Clay County Sesquicentennial (1822–1972) Souvenir Program. Liberty, MO: Al's Printing Service, 1972. WMA Collection, WMSHS.

Cole, Arthur Harrison. *The American Wool Manufacture.* 2 vols. Cambridge, MA: Harvard University Press, 1926.

Cooper, Grace Rogers. "Watkins Mill Revisited." *Museum News* (September 1963): 11–15.

Cooper, James C., and Karl Borden. "The Interpretation of Wages and Prices in Public Historical Displays." *The Public Historian* 19 (Spring 1997): 9–29.

Cousins, Peter H. "Tall Timber, Wheat and Wrinkly Sheep: Three Generations of Ohio Farmers." *Henry Ford Museum and Greenfield Village Herald* 14 (1985): 3–27.

Cowan, Ruth Schwartz. *A Social History of American Technology.* New York: Oxford University Press, 1992.

Craven, Morgan. Interview by Ann Matthews (Sligar), 1981. Typescript. WMSHS.

Crockett, Norman. "Frontier Dentist: The Letters of Waltus Jewell Watkins." *The Colorado Magazine* 44 (Fall 1967): 277–92.

———. "The Westward Movement and the Transit of American Machine Technology: The Case of Wool Manufacturing." *Nebraska Journal of Economics and Business* 8 (Summer 1969): 111–21.

———. *The Woolen Industry of the Midwest.* Lexington: University Press of Kentucky, 1970.

Cummins, Jim. *Jim Cummins' Book.* Denver: Reed, 1903.

Dagley, Charles. Interview by Booker H. Rucker, c. 1965. Transcript of notes. WMSHS.

Dains, Mary K. "Alexander Campbell and the Missouri Disciples of Christ." *Missouri Historical Review* 77 (October 1982): 13–46.

Dalton, John M., Papers. Western Historical Manuscripts Collection, University of Missouri-Kansas City.

Daughters of the American Revolution (DAR), Missouri Society, Alexander Doniphan Chapter. *Clay County, Missouri: Centennial Souvenir, 1822–1922.* Liberty, MO: *Liberty Tribune,* 1922.

De Voto, Bernard. *The Year of Decision, 1846.* Boston: Houghton Mifflin, 1989.

Doyle, Don Harrison. *The Social Order of a Frontier Community: Jacksonville, Illinois 1825–70.* Urbana: University of Illinois Press, 1978.

Dunwell, Steve. *The Run of the Mill: A Pictorial Narrative of the Expansion, Dominion, Decline, and Enduring Impact of the New England Textile Industry.* Boston: Godine, 1978.

Dye, Lee. "When Eggs are Dear: Nineteenth Century Foodways and Domesticity." Unpublished paper, University of Missouri-Kansas City, 1996. WMSHS.

Ellis, David M., ed. *The Frontier in American Development: Essays in Honor of Paul Wallace Gates.* Ithaca, NY: Cornell University Press, 1969.

Emerson, Ralph Waldo. "Works and Days." In *The Complete Works of Ralph Waldo Emerson.* Vol. 7, *Society and Solitude,* 1904. Reprint, New York: AMS Press, 1979.

Entriken, Bertha Bell. *Smokey Row.* Cameron, MO: Smokey Row, Inc., 1963.

Etcheson, Nicole. *The Emerging Midwest: Upland Southerners and the Political Culture of the Old Northwest 1787–186.* Bloomington: Indiana University Press, 1996.

Evans, Clement A., ed. *Confederate Military History: A Library of Confederate States History....* Vol. 9. Atlanta, GA: Confederate Publishing, 1899.

Evans, Priscilla Ann, "Merchant Gristmills and Communities, 1820–1880: An Economic Relationship." *Missouri Historical Review* 68 (April 1974): 317–26.

Everett, Edward. "On the Importance of Scientific Knowledge to the Practical Men, and on the Encouragements to its Pursuit," in *Orations and Speeches on Various Occasions,* Vol. 1, 8th ed. (Boston: Little, Brown and Co., 1870).

Excelsior Springs (MO) Daily Standard, 1945.

Excelsior Springs (MO) Shopper, 1960.

Excelsior Springs (MO) Town and Country Leader, 1962.

Faragher, John Mack. *Sugar Creek: Life on the Illinois Prairie.* New Haven, CT: Yale University Press, 1986.

Feller, Daniel. *The Public Lands in Jacksonian Politics.* Madison: University of Wisconsin Press, 1984.

Fellman, Michael. *Inside War: The Guerrilla Conflict in Missouri During the American Civil War.* New York: Oxford University Press, 1989.

Flader, Susan, ed. *Exploring Missouri's Legacy: State Parks and Historic Sites.* Columbia: University of Missouri Press, 1992.

Fowler, Orson S. *The Octagon House, A Home for All.* 1853. Reprint, New York: Dover, 1973.

Goldfield, David, Carl Abbott, Virginia Dejohn Anderson, Jo Ann E. Argersinger, Peter M. Argersinger, William L. Barney, and Robert M. Weir. *The American Journey: A History of the United States.* Combined 2nd ed. Upper Saddle River, NJ: Prentice Hall, 2001.

Goodrich, Thomas. *Black Flag: Guerrilla Warfare on the Western Border, 1861–1865.* Bloomington: Indiana University Press, 1995.

Goodwin, Will. Interview by Booker H. Rucker, c. 1964. Typescript. WMSHS.

Gordon, Robert B., and Patrick M. Malone. *The Texture of Industry: An Archaeological View of the Industrialization of North America.* New York: Oxford University Press, 1994.

Gross, Laurence. Audiotape recordings on mill machinery, June 1978. WMSHS.

——. *The Course of Industrial Decline: The Boott Cotton Mills of Lowell, Massachusetts, 1835–1955.* Baltimore, MD: Johns Hopkins University Press, 1993.

——. "The Importance of Research Outside the Library: Watkins Mill, A Case Study." *IA: The Journal of the Society for Industrial Archaeology* 7 (1981): 15–26.

——. *Watkins Mill: A Report on Equipment and Procedures for the Historic American Engineering Record.* National Park Service, Historic American Engineering Record Program. September 1978. Report no. HAER-MO-1, part 1.

——. "Wool Carding: A Study of Skills and Technology." *Technology and Culture* 28 (October 1987): 804–27.

Gutman, Herbert. *Work, Culture and Society in Industrializing America.* New York: Vintage Books, 1976.

Hall, Jacquelyn Dowd, Robert Korstad, and James Leloudis, "Cotton Mill People: Work, Community and Protest in the Textile South, 1880–1940." *American Historical Review* 91 (June 1986): 245–86.

Hall, Jacquelyn Dowd, James Leloudis, Robert Korstad, Mary Murphy, LuAnn Jones, Christopher B. Daly, and Michael Frisch. *Like A Family: The Making of a Southern Cotton Mill World.* Chapel Hill: University of North Carolina Press, 1987.

Hall, Kermit. *The Magic Mirror: Law in American History.* New York: Oxford University Press, 1989.

Handy, Robert T. *A History of the Churches in the United States and Canada.* New York: Oxford University Press, 1977.

Henderson, N. Clifford. "A Proposal for the Reconstruction of the Smithy at Watkins Mill State Historic Site." Master's thesis, University of Missouri–Kansas City, 1992.

Henretta, James A., W. Elliott Brownlee, David Brody, and Susan Ware. *America's History to 1877.* New York: Worth, 1997.

Hindle, Brooke, and Steven Lubar. *Engines of Change: The American Industrial Revolution 1790–1860.* Washington, DC: Smithsonian Institution Press, 1986.

Hippen, James C. "Industrial Textile Machinery: Five North American Museums." *Technology and Culture* 10 (October 1969), 574–86.

History of Clay and Platte Counties, Missouri. St. Louis: National Historical Company, 1885.

Homespun to Factory Made: Woolen Textiles in America 1776–1876. North Andover, MA: Merrimack Valley Textile Museum, 1977.

Hood, Adrienne D. *The Weaver's Craft: Cloth, Commerce, and Industry in Early Pennsylvania* (Philadelphia: University of Pennsylvania Press, 2003).

Howe, Daniel Walker. "The Market Revolution and the Shaping of Identity in Whig-Jacksonian America." In *The Market Revolution in America: Social, Political and Religious Expressions 1800–1880,* edited by Melvyn Stokes and Stephen Conway, 259–81. Charlottesville: University Press of Virginia, 1996.

Hurt, R. Douglas. *Agriculture and Slavery in Missouri's Little Dixie*. Columbia: University of Missouri Press, 1992.

——. "Planters and Slavery in Little Dixie." *Missouri Historical Review* 88 (July 1994): 397–415.

Hyman, Harold M., and William M. Wiecek. *Equal Justice Under Law: Constitutional Development 1835–1875*. New York: Harper & Row, 1982.

An Illustrated Historical Atlas of Clay County, Missouri. 1877. Reprint, Liberty, MO: Clay County Historical Society, ca. 1977.

Jackson, John Brinckerhoff. *Discovering the Vernacular Landscape*. New Haven, CT: Yale University Press, 1984.

Jefferson, Thomas. Letter to James Madison, 27 April 1809. Library of Congress, http://www.loc.gov/exhibits/jefferson/149.html (accessed March 2003).

Kammen, Michael. *Mystic Chords of Memory: The Transformation of Tradition in American Culture*. New York: Knopf, 1991.

Kansas City (MO) Star, 1923–97 (noninclusive).

Kansas City (MO) Times, 1942, 1953, 1962–63.

Kasson, John F. *Civilizing the Machine: Technology and Republican Values in America, 1776–1900*. New York: Penguin Books, 1977.

Klinkenborg, Verlyn. "If It Weren't for the Ox, We Wouldn't Be Where We Are." *Smithsonian* 24 (September 1993): 82–83.

Kohl, Lawrence Frederick. "The Concept of Social Control and the History of Jacksonian America." *Journal of the Early Republic* 5 (Spring 1985): 21–34.

——. *The Politics of Individualism: Parties and the American Character in the Jacksonian Era.* New York: Oxford University Press, 1989.

Larkin, Jack. *The Reshaping of Everyday Life 1790–1840*. New York: Harper and Row, 1988.

Launius, Roger D. *Alexander William Doniphan: Portrait of a Missouri Moderate*. Columbia: University of Missouri Press, 1997.

Lee, Judith Yaross. "Paradise Limited: An Interview with Leo Marx." *American Heritage of Invention & Technology* 4 (Fall 1988): 34–39.

Lemmer, George Francis. "Agricultural Improvement in Missouri: 1830 to the Civil War." Master's thesis, University of Missouri–Columbia, 1941.

Lentz, Elizabeth S., and Laurel E. Wilson. "Mill Girls on the Missouri Frontier: The Impact of Wage Workers on their Family Economy." Unpublished paper, University of Missouri–Columbia, 1996. WMSHS.

——. "The Mill Ran Today: Patterns of Work in a Mid-Nineteenth Century Missouri Woolen Mill." Unpublished paper, University of Missouri–Columbia, c. 1997. WMSHS.

Liberty (MO) Advance, 1939.

Liberty (MO) Chronicle, 1934.

Liberty (MO) Tribune, 1855–84, 1953, 1971–78. WMA Collection, WMSHS.

Licht, Walter. *Industrializing America: The Nineteenth Century*. Baltimore: Johns Hopkins University Press, 1995.

Maier, Pauline, Merritt Roe Smith, Alexander Keyssar, and Daniel J. Kevles. *Inventing America: A History of the United States.* New York: Norton, 2002.

Marx, Leo. *The Machine in the Garden: Technology and the Pastoral Idea in America.* New York: Oxford University Press, 1964.

Matson, Madeline. *Food in Missouri: A Cultural Stew.* Columbia: University of Missouri Press, 1994.

Matthews (Sligar), Ann. "The Building as a Document: Identifying the Worker's Habits." Report presented at the Annual Meeting of the Society for Industrial Archaeology, Saint Paul, MN, May 1983. WMSHS.

——. "An Engineering Treasure is Restored to Its Original Industrial Splendor." *National Engineer* (October, 1982): 13–15.

——. Interview by Patricia Needham, 13 February 1987. Audiotape recording. Western Historical Manuscripts Collection, University of Missouri–Kansas City.

——. Interviews by. See Craven, Morgan.

——. "Why We Decided Not to Demonstrate Process at Watkins Woolen Mill." In Proceedings of the 1986 ALHFAM Annual Conference, 67–72. Association of Living Historical Farms and Agricultural Museums, 1989.

——. "The Woolen Trade in Late 19th Century Missouri." *Midwest Open-Air Museums Magazine* 10 (Spring 1989): 8–14.

McCandless, Perry. *History of Missouri, 1820 to 1860.* Vol. 2 of *History of Missouri,* edited by William E. Parrish. Columbia: University of Missouri Press, 1972.

McGraw, Judith. "'So Much Depends Upon a Red Wheelbarrow': Agricultural Tool Ownership in the Eighteenth-Century Mid-Atlantic." In Smith and Clancey, *Major Problems,* 88–101.

McMillen, Margot Ford. *Paris, Tightwad, and Peculiar: Missouri Place Names.* Columbia: University of Missouri Press, 1994.

Means, Chad. "The Edited Watkins Family Letters, with General Introduction, Special Introductions, and Explanatory Footnotes." Unpublished paper, William Jewell College, 1991. WMSHS.

Meinig, Donald W. *The Shaping of America: A Geographical Perspective on 500 Years of History.* Vol. 2, *Continental America 1800–1867.* New Haven, CT: Yale University Press, 1993.

Meyer, Duane G. *The Heritage of Missouri.* Springfield, MO: Emden Press, 1982.

Missouri: A Guide to the "Show Me" State. New York: Duell, Sloan & Pierce, 1941.

Norris, James D. *R. G. Dun & Co., 1841–1900: The Development of Credit-Reporting in the Nineteenth Century.* Westport, CT: Greenwood Press, 1978.

North Kansas City (MO) Dispatch, 1962.

Oberholtz, L. E. "Early Pioneering of Waltus L. Watkins." In *Spinning Yarns.*

——. Interview by Patricia Needham, 27 February 1987. Audiotape recording. Western Historical Manuscripts Collection, University of Missouri–Kansas City.

——. Interviews by Louis Potts, 23, 25, and 30 June 1987, 29 February 1988, and 15 April 1988. Audiotape recordings. Western Historical Manuscripts Collection, University of Missouri–Kansas City.

——. "Life Span of Waltus L. Watkins." In *Spinning Yarns.*

O'Brien, Michael. *Grassland, Forest, and Historical Settlement: An Analysis of Dynamics in Northeast Missouri.* With contributions by Jacqueline A. Ferguson et al. Lincoln: University of Nebraska Press, 1984.

Ogren, Paul. *Watkins Mill: An Economical Business History of Watkins Mill.* National Park Service, Historic American Engineering Record Program. [1978]. Report no. HAER-MO-1, part 2.

"Old Men of Clay County." *Liberty (MO) Tribune,* 11 November 1870.

Parrish, William E. *A History of Missouri, 1860 to 1875.* Vol. 3 of *History of Missouri,* edited by William E. Parrish. Columbia: University of Missouri Press, 1973.

——. *Missouri Under Radical Rule, 1865–1870.* Columbia: University of Missouri Press, 1965.

Petty, Evelyn. "Old Clay is Some Punkins…A History of Clay County." *Liberty (MO) Tribune,* 18 November 1971–8 February 1978. Clay County Archives, Liberty, MO.

Portrait and Biographical Record of Clay, Ray, Carroll, Chariton and Linn Counties, Missouri. Chicago: Chapman Brothers, 1893, s.v. "Waltus L. Watkins."

Potts, Louis W. "The Franklin Debate Society: Culture on the Missouri Frontier." *Missouri Historical Review* 86 (October 1991): 1–21.

——. "From Speculator to Gentleman: Waltus L. Watkins and the Case for Land Speculation on the Missouri Frontier." *Gateway Heritage* 13 (Fall 1992): 46–55.

——. Interviews by. See Oberholtz, L. E.; Rucker, Booker H.

——. "Recollections and Reconstruction in Clay County, Missouri, 1870." *Kawsmouth* 4 (Summer–Autumn 2002), 68–83.

——. "Waves of Revivalism in Clay County, 1840–1918." *Missouri Historical Review* 88 (April 1994): 262–78.

Prude, Jonathan. *The Coming of Industrial Order: Town and Factory Life in Rural Massachusetts 1810–1860.* Cambridge: Cambridge University Press, 1983.

Pusateri, C. Joseph. *A History of American Business.* 2nd ed. Arlington Heights, IL: Harlan Davidson, 1988.

Rohrbaugh, Malcolm J. *The Land Office Business: The Settlement and Administration of American Public Lands, 1789–1837.* New York: Oxford University Press, 1968.

Rold, William. "Secessionist Strength in Missouri." *Missouri Historical Review* 72 (July 1978): 412–23.

Roney, Ruth Bogart. "Advice to the Young Pioneer—W.L. Watkins." In *Spinning Yarns.*

——. "Data on the Watkins Woolen Mill," 1953. WMA Collection, WMSHS.

——. "Early Pioneering of Waltus L. Watkins." In *Spinning Yarns.*

——. *From Entry Fee to 'Fifty-Three': History of Lawson's Three County Community.* Clinton, MO: The Printery, 1953, 1970.

——. "Historic Names in Tale of Old Church," *Kansas City Star*, 28 February 1961.

——. "Old Watkins Mill: A Picturesque West Missouri Landmark." Draft manuscript, c. 1962. WMSHS.

Rorabaugh, W. J. *The Alcoholic Republic, An American Tradition.* New York: Oxford University Press, 1979.

Rose, Anne C. *Voices of the Marketplace: American Thought and Culture.* New York: Twayne, 1995.

Rucker, Booker H. "Archaeological Investigations at Watkins Mill State Park, Lawson, Missouri." Unpublished report to Missouri State Park Board, January 1964. WMSHS.

——. Interview by Louis Potts, 21 September 1988. Audiotape recording. Western Historical Manuscripts Collection, University of Missouri–Kansas City.

——. Interviews by. See Dagley, Charles; Goodwin, Will.

——. "Watkins Mill." *The Harbinger Magazine* (December 1970–January 1971): 18–19.

——. "Watkins Mill in Transition: The Death of Waltus Watkins." In *Spinning Yarns.*

Russell, Janet A. "Watkins Mill Association." Unpublished paper, University of Missouri–Kansas City, 1996. WMSHS.

"School Days." *Historic Preservation* 11 (1959): 100–3.

Schroeder, Walter A. *Presettlement Prairie of Missouri.* Natural History Series no. 2 (revised). Jefferson City: Missouri Department of Conservation, 1982.

Scranton, Philip. "Varieties of Paternalism: Industrial Structures and the Social Relations of Production in American Textiles." *American Quarterly* 36 (Summer 1989): 235–57.

Sellers, Charles. *The Market Revolution: Jacksonian America, 1815–1846.* New York: Oxford University Press, 1991.

Settle, William A., Jr. *Jesse James Was His Name; or, Fact and Fiction Concerning the Career of the Notorious James Brothers of Missouri.* Columbia: University of Missouri Press, 1966.

Slates, B. Y. "Watkins Mill Steam Engine." Information sheet, n.d. WMSHS.

Sligar, Ann Matthews. *Waltus Watkins and His Mill.* Lawson, MO: Watkins Mill Association, 2002.

——. See also Matthews (Sligar), Ann.

Smith, Merritt Roe. "Industry, Technology, and the 'Labor Question' in 19th Century America: Seeking Synthesis." *Technology and Culture* 32 (July 1991): 556–62.

Smith, Merritt Roe, and Gregory Clancey, eds. *Major Problems in the History of American Technology.* Boston: Houghton Mifflin, 1998.

Smith, Rogers M. *Civic Ideals: Conflicting Visions of Citizenship in U.S. History.* New Haven, CT: Yale University Press, 1997.

Society for the History of Technology (SHOT). "Whole Cloth: Discovering Science and Technology Through American History." Jerome and Dorothy Lemelson Center for the Study of Invention and Innovation, Smithsonian Institution, Washington, D.C, http://www.si.edu/lemelson/centerpieces/whole_cloth/ (accessed February 2004).

Spinning Yarns. Watkins Mill Association publication, n.d. WMA Collection, WMSHS.

Statistical History of the United States from Colonial Times to the Present. Stamford, CT: Fairfield, 1965.

Stiles, T. J. *Jesse James: Last Rebel of the Civil War.* New York: Knopf, 2002.

Thelen, David. *Paths of Resistance: Tradition and Dignity in Industrializing Missouri.* New York: Oxford University Press, 1986.

Thorp, Joseph. *Early Days in the West: Along the Missouri One Hundred Years Ago.* Liberty, MO: Liberty Tribune, 1924.

Ulrich, Laurel Thatcher. *The Age of Homespun: Objects and Stories in the Creation of an American Myth.* New York: Knopf, 2001.

United States Biographical Dictionary and Portrait Gallery of Eminent and Self-Made Men. Missouri vol. New York: United States Biographical Publishing, 1876, s.v. "Waltus Lockett Watkins."

United States Census Bureau. Population, Slave, Industrial, and Agricultural Schedules. Fifth–Twelfth Censuses, Washington Township, Clay County, MO, 1830–1900. WMSHS.

United States Census Office. *Statistics of the United States of America, as Collected and Returned... Under the Thirteenth Section of the Act for Taking the Sixth Census....* Washington, DC: Blair and Rives, 1841.

Voss, Stuart F. "Town Growth in Central Missouri, 1815–1860." Pts. 1–3. *Missouri Historical Review* 64 (October 1969): 64–80; (January 1970): 197–217; (April 1970): 322–50.

Wallace, Michael. "Reflections on the History of Historic Preservation." In Benson, Brier, and Rosenzweig, *Presenting the Past,* 165–99.

The War of Rebellion: A Compilation of the Official Records of the Union and the Confederate Army. c. 1900. Reprint, Harrisburg, PA: National Historical Society, 1971.

Watkins, J. Spencer. *Lucky Montana Cowpoke.* New York: Vantage Press, 1958.

"Watkins Mill," Historic American Buildings Survey/Historic American Engineering Record (HABS/HAER), American Memory Collections, Library of Congress, Washington, D.C., http://lcweb2.loc.gov/ammem/amhome.html (accessed 19 February 2004).

Watkins Mill State Historic Site (WMSHS), Lawson, MO. Includes Watkins Family Collection, Watkins Mill Association (WMA) Collection, minute books of the Mount Vernon Missionary Baptist Church and the Franklin Debate Society, Ray Coffman Papers, WMSHS and WMA Research Files.

Welch, Debbie K. "Silent Testimony: A Material Culture Reading of the 1854 Residence of Waltus and Mary Ann Watkins." Unpublished paper, University of Missouri–Kansas City, 1992. WMSHS.

White, Richard. *The Middle Ground: Indians, Empires, and Republics in the Great Lakes Region, 1650–1815.* Cambridge: Cambridge University Press, 1991.

Wiebe, Robert H. *The Opening of American Society: From the Adoption of the Constitution to the Eve of Disunion.* New York: Vintage Books, 1984.

Wingerd, Mary Lethert. "Rethinking Paternalism: Power and Parochialism in a South Mill Village." *Journal of American History* 83 (December 1996): 872–902.

Wood, Gordon S. "The Enemy is Us: Democratic Capitalism in the Early Republic." *Journal of the Early Republic* 16 (Summer 1996): 293–308.

Woodson, W. H. *History of Clay County.* Topeka, KS: Historical Publishing Company, 1920.

Yarbrough, Jean C. *American Virtues: Thomas Jefferson on the Character of a Free People.* Lawrence: University Press of Kansas, 1998.

Young, Mary E. "Congress Looks West: Liberal Ideology and Public Land Policy in the Nineteenth Century." In *The Frontier in American Development: Essays in Honor of Paul Wallace Gates,* edited by David M. Ellis, 74–116. Ithaca, NY: Cornell University Press, 1969.

Index

italic locators = illustrations in text
c- locators = illustrations in color insert